m Z C P A

ad

Q

datis

B

D

...ta APCZ horizonti parallela
...n Cyclois quaecunque AQP rectae AB
...ns in puncto Q, item Cyclois alia
...prioris basem et altitudinem re-
... Cyclois novissima transibit per
...neque in qua grave a puncto A
...cibissime perveniet. Q. E. I.

PHYSICS

AN ILLUSTRATED HISTORY OF THE
FOUNDATIONS OF SCIENCE

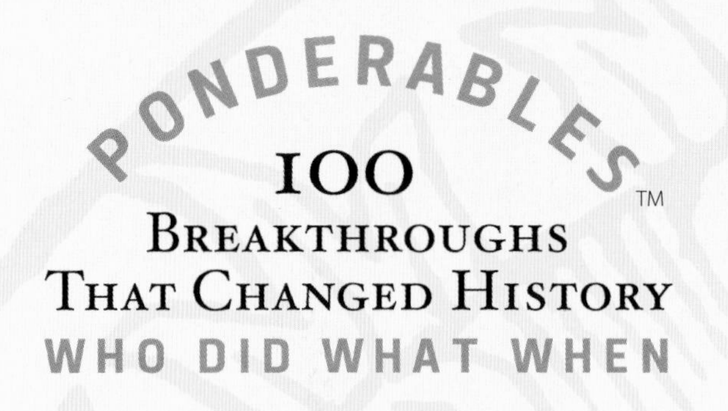
PONDERABLES™
100
Breakthroughs
That Changed History
WHO DID WHAT WHEN

PHYSICS

AN ILLUSTRATED HISTORY OF THE FOUNDATIONS OF SCIENCE

Tom Jackson

SHELTER HARBOR PRESS
NEW YORK

Contents

THE SUBATOMIC AGE

MODERN PHYSICS

Introduction

PHYSICS IS THE FOUNDATION OF ALL SCIENCE. WITHOUT IT ALL OF OUR OTHER KNOWLEDGE WOULD CRUMBLE AND COLLAPSE. WE CAN NOW STUDY NATURE AT THE SMALLEST SCALES, BUT THERE IS A LOT THAT THIS SCIENCE HAS STILL TO DISCOVER.

A PONDERABLE

The thoughts and deeds of great thinkers always make great stories, and here we have one hundred all together. Each story relates a ponderable, a weighty problem that became a discovery and changed the way we understand the world and our place in it.

Knowledge does not arrive fully formed. We have to work at it, taking it in turns to consider the evidence and add our take on it. In hindsight even the most cutting-edge opinions can look utterly wrong, if not bizarre and laughable. But our high-tech, inter-connected world is built on these ponderables, growing, changing step-by-step into an ever clearer picture of reality.

THE NATURE OF THINGS

The story of physics is simply an enquiry into nature. Even the word *physics* means "nature" in ancient Greek. Since the beginning of civilization, humanity has wondered what gives the air, water, and

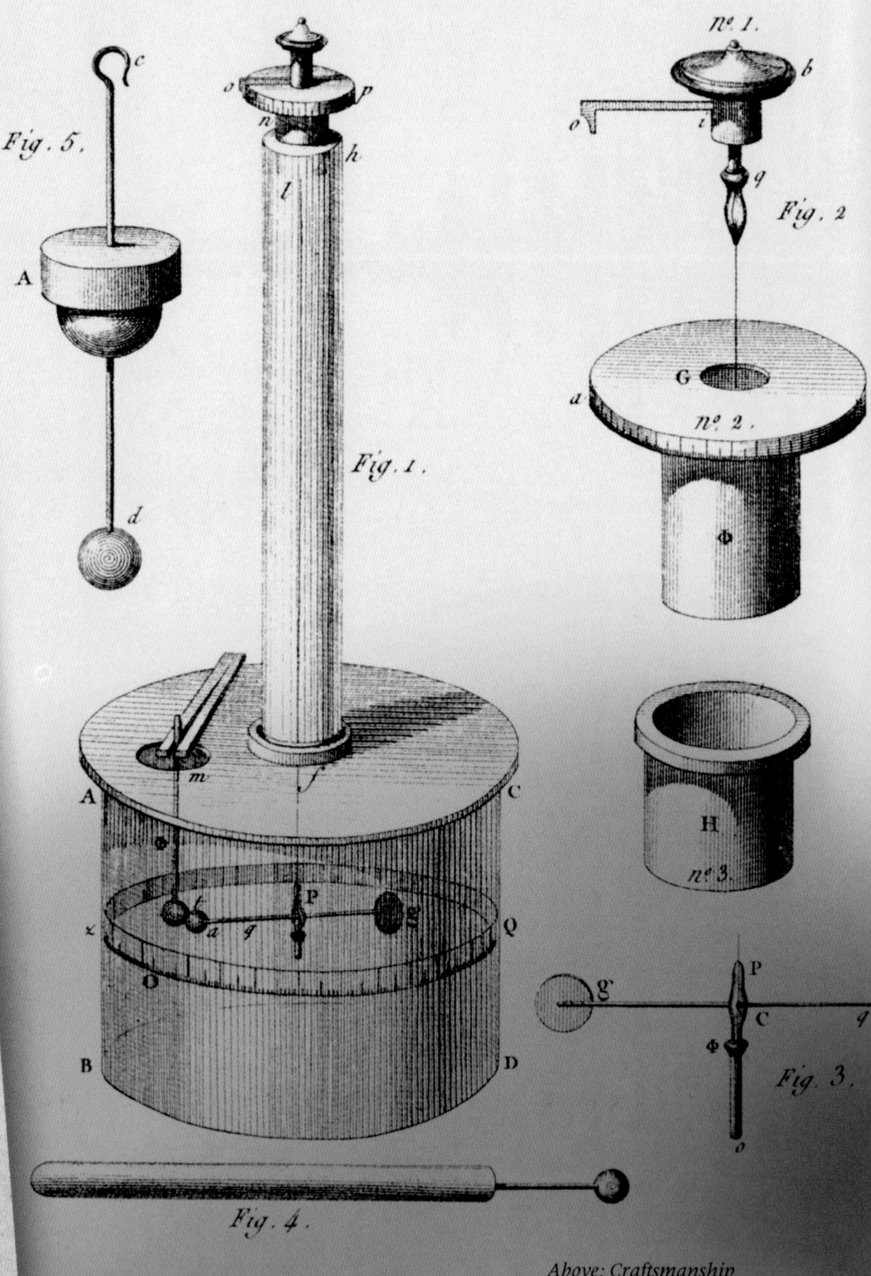

Left: A 16th-century Arabic manuscript contains a design for machinery that converts the flow of water into a useful rotational motion.

Above: Craftsmanship meets precision in the sensitive torsion balance used to measure the strength of electrical charge.

earth its form and how they related to the stars glinting from beyond our world. The Earthly realm appeared to be in constant flux, always changing, while the heavens appeared—at first—to be fixed.

The seed from which physics bloomed is the idea that a single set of laws governs everything in the Universe, from the churning of the ocean's waves to the reflections of starlight on its waters. Centuries of thought and experimentation—plus many chance discoveries—have shown this intuition to be correct: The universal laws of physics allow us to understand the behavior of a star in a distant galaxy just as well as the movement of electricity through a supercomputer. This incredible scope is what makes physics the ultimate science.

DAUGHTERS OF PHYSICS

Name a science: Chemistry, biology, geology? The list could be much longer, and each subject has its own proud history and makes immeasurably significant contributions to our compendium of knowledge. However, they all rely on physics to form a platform from which they reach for new understanding.

Chemistry explains the structure of the millions of substances, both natural and artificial, that make up our world. To do that, it takes the atom as described by physics and studies how it can be involved in chemical reactions, making and breaking bonds, to form the many materials found on Earth. Biology reports on how life-forms—the Universe's most complex systems—function. It relies on chemistry to track the flow of energy that makes a body alive. Geology tells us that even solid ground is on the move. To do so it has deployed an understanding of physics, namely the study of heat, acoustics, and materials, to show how great forces inside the planet and outside it are constantly reshaping the surface.

Let's take a look at how, physics, this most powerful of disciplines, emerged from myths and guesswork to become the foundation of all scientific knowledge.

Right: Highly engineered refrigerators capable of the lowest temperatures possible led to the discovery of superconductors—which levitate in a magnetic field.

Above: Modern particle accelerators and sensitive detectors allow physicists to probe the ultrafine structure of the very smallest subatomic particles.

BRANCHES OF PHYSICS

As with all scientific disciplines, physics is divided into a series of branches, each specializing in a particular area of study. However, unlike other sciences, the branches of physics are divided into two distinct groups—classical and modern. As the names suggest, classical physics was developed before modern physics. In most sciences, older ideas are replaced by new ones, but modern physics is based on a different set of principles to its classical bedfellow, and so both remain valid, although unconnected, areas of study.

CLASSICAL MECHANICS

The study of how objects of different mass move due to the application of force.

ELECTROMAGNETISM

The study of electrical charge, electric currents, magnetism, and the spectrum of electromagnetic radiation, which includes visible light, radio waves, and X rays.

STATISTICAL MECHANICS

A mathematical technique for modeling the motion of invisible molecules and atoms.

CLASSICAL PHYSICS

This is old-school physics. It deals with phenomena perceived on the human scale, such as the motion of an object, energy transfers in machinery, or the production of electricity or sound. In the late 19th century, it was believed by some that classical theories had solved every mystery that science had to offer. By 1910, they were proved very wrong with the advent of modern theories.

Below: A device for measuring the link between heat and work.

ACOUSTICS

The study of sound waves that travel through physical media.

OPTICS

The study of the nature of light and the behavior of light rays.

THERMODYNAMICS

The study of how heat is transferred through materials and converted into other forms of energy, such as motion or light.

PROPERTIES OF MATTER

Understanding how different materials have different properties.

RELATIVITY

How a mass interacts with space and time as it moves.

QUANTUM MECHANICS

The study of physical phenomena at the level of the smallest subatomic particles.

Below: An explosion caused by nuclear fusion.

NUCLEAR PHYSICS

Studying the structure and behavior of the atom.

MODERN PHYSICS

As physics matured in the early 20th century, classical theories were found to "break down" when applied to the extremes, where even tiny errors became significant. The modern theory of relativity was developed to tackle space and time on the large scale, while quantum theory studies the nature of matter at the very smallest scale. The great hope of 21st-century physics is to be able to combine these two theories.

CONDENSED MATTER PHYSICS

Understanding liquids and solids in terms of how the atoms and molecules they contain are operating at the quantum level.

PARTICLE PHYSICS

Describing the behavior of fundamental particles that make up matter and the forces acting between them.

ASTROPHYSICS AND COSMOLOGY

A shared field with astronomy that uses atomic physics and other modern theories to explain the behavior of stars and formation of the Universe.

THE DAWN OF SCIENCE

1 Explaining Nature

AS WE HAVE SAID ALREADY—AND WILL NO DOUBT SAY AGAIN—THE WORD *PHYSICS* COMES FROM THE GREEK FOR "NATURE." The human ability to investigate nature is itself natural.

The Garden of Eden according to Hieronymus Bosch. The Fall, described in the Eden story, when Adam and Eve first feel shame and embarrassment is said by some to indicate the moment humanity achieved a "theory of mind." This is the concept that we all share, where we know that our thoughts are different to the contents of the minds of those around us. Do the other animals of Creation have a theory of mind? We think not, but they might be thinking that about us.

Like any creature that must eat or be eaten, the primitive human being would have been constantly on the lookout. We observe the surroundings in great detail and come to some conclusions about what is likely to happen next. To do that we rely on our past experiences—what happened last time—but we are also able to think laterally, taking unrelated knowledge and applying it to a new situation. Put simply we can imagine things that have not happened yet, may never happen, or cannot happen.

Decisions, decisions

Humans have a primate brain, which has a capacity to make a large number of decisions quickly while also being perpetually interested in novelty. (Being curious helps you find what you need, perhaps before you know you need it.) The human brain is especially well developed, able to create mental maps of an area, including how it changes through the seasons. But above all we use our gray matter to cooperate with our fellow humans and gather information from them that helps us survive. If you make a mistake—and live to tell the tale—you won't make it twice. Humans can share knowledge of successes and failures with each other, so we do not merely learn by experience but are also taught by the experiences of others.

THE FLAWED OLYMPIANS

In general, creation relies on all-powerful deities who also have all the answers. Questioning the veracity of these stories is to question the gods themselves. However, ancient Greeks worshiped a bunch of divine beings who lived on Mount Olympus (a real mountain). The Olympian gods were very human, frequently falling in love or fighting, and did not seem to be entirely in control. It was against this backdrop that the early Greek philosophers were able to ask the big questions about the Universe—the rest is the history of science.

The all-too-human Olympian gods pose for a family portrait.

Questions, questions

This cumulative wisdom passed down through generations leads to culture. This is a toolkit of knowledge and traditions that provide the answers for many important questions, such as where can we find food this time of year? How long until the crops ripen? When will the river flood? But there are some questions that mere

Creation stories from India and China often refer to the world, sky, or heavens being supported on various animals—normally mighty elephants, long flexible serpents, or sturdy turtles.

experience cannot answer: Where did nature come from?

Creating understanding

To complete the picture, we humans did what we always do: We imagined a possible answer. Our cultural tools became stitched into a complete world view. This still contained the necessary components to ensure we avoided starvation, but also gave humans a place in nature—and explained where it came from.

There are as many of these creation myths as there are cultures. The Boshongo people of Central Africa believe the Universe was formed from the vomit of the great Bumba. Other myths explain that the world is the offspring of a mother (and perhaps father) nature, others see it as an ordered nature forming from chaos, while the most famous ones describe how it was all created from nothing at all. There is no solid evidence for any of it, which is where the story of physics begins.

In one sense physics is a creation story like any other. Currently, it asserts that the Universe formed from nothing. The difference is that every bit of the physics story is based on tested evidence—and any of it could change at any time. Let's look at what the story tells us right now.

THE PIRAHÃ

In 1980, American anthropologist Daniel Everett described the mythology of the Pirahã people of Brazil: There is none. The Pirahã only believe what they experience themselves and will not accept anything they are told unless the teller experienced the event his or herself. They make no effort to accumulate knowledge and simply purchase innovations from neighboring tribes.

2 Thales, the Father

If science begins anywhere, it begins with physics, and if physics begins anywhere it is in ancient Greece, specifically the city of Miletus. This was home to Thales, traditionally viewed today as the founding father of science as we know it.

While his deeds had a big impact, little is known about Thales, the man.

With a life that crossed from the 7th to 6th centuries BCE, Thales is by no means a straightforward historical figure. None of his writings have survived, but the work of many later philosophers leads inexorably back to Thales. He was a Greek in the sense of belonging to the ancient Hellenic culture; his home city was on the western cost of what is now Turkey. That was a hot spot on trade routes of the time, and so Thales would certainly have been exposed to the older civilizations of Egypt and Babylon—perhaps even traveling there himself.

Thales can be called a scientist only in retrospect. He was the first person on record to eschew mystical reasoning and puzzle out the causes of natural phenomena by only what could be observed about them. Science as we know it would take another 20 centuries to develop, and through the lens of modern understanding, Thales's big theory of nature appears almost childlike: First he had to decide what the Universe was made from. Thales was a monist, and believed that all things were made from one substance: Water. His reasoning was that only water has three unique features: It is essential for life, it can move and flow, and it can change its form.

Thales had a more lasting impact with his geometry (there is a theory of triangles named for him), and his ability to predict solar eclipses is said to have helped end a war between local kingdoms.

THE SEVEN SAGES

Thales is one of the Seven Sages of Greece, philosophers, politicians and jurists from the 7th and 6th centuries BCE who are said to have sown the seeds of the modern world: The remaining six are Cleobulus of Lindos, Solon of Athens, Chilon of Sparta, Bias of Priene, Pittacus of Mytilene, and Periander of Corinth.

A 19th-century montage, based on various bearded busts, shows the seven sages (and a few guests) gathered at an imaginary banquet. The most likely Thales figure is seated far right.

3 Atoms: Starting Small

MANY MODERN PHYSICISTS DEVOTE THEIR LIVES TO CATCHING A GLIMPSE OF WHAT GOES ON INSIDE ATOMS. Such a thought would certainly have made Democritus, the Greek philosopher who outlined the nature of atoms 2,400 years ago, give a broad grin. After all, he was right—almost.

LEUCIPPUS

The concept of the atom was first proposed by Democritus's teacher, Leucippus, who died while his student was still a boy. He had originally suggested that atoms moved in a random, indeterminable way. Democritus took the opposite view: All motion, form, and change were the result of specific interactions of atoms.

Monists, the followers of Thales, were tricky customers. For example, a chat with Parminedes from the 5th century could have gone something like this: "Everything in the Universe is made of one 'thing,' and so it is impossible to have 'nothing.' For a 'thing' to move it must occupy a place previously filled by 'nothing'—which, as I've said, is impossible. So all the movement and change we see around us is an illusion."

Small change

A new theory of atomism was developed to rebut this logical merry-go-round, first by Leucippus but developed fully by his pupil Democritus, who lived in Thrace.

Atomism stated that matter ("things") could not be divided up indefinitely. Eventually, you got down to atoms—minute, invisible, and indivisible building blocks of all things. (The word *atom* is derived from the ancient Greek for "indivisible.") Any changes in nature were merely due to atoms being rearranged. The idea that matter had more than a single ingredient had not been completely killed off by monism, and Democritus suggested that atoms need not be identical. Large scale characteristics of matter could be explained by the small scale features of their atoms. Liquid water, for example, was comprised of smooth, rounded atoms that flowed past each other easily. Solid materials were made from hooked atoms that clung together—salt atoms were especially spiky, which gave them their sharp taste. Even the soul was made of atoms; they were so small they could pass through solid matter. However, without proof, atoms would remain just an idea until the 1800s.

Democritus is remembered as a cheerful soul, the Laughing Philosopher. This 1628 portrait by Dutch master Hendrik ter Brugghen has something of the carouser about it.

4 Four Elements and More

THE MONIST PRINCIPLE OF THALES, THAT ALL IS WATER, WAS NOT ENOUGH FOR THE PHILOSOPHERS THAT CAME IN HIS WAKE. Instead, they relied on more basic, primitive intuition that the Universe was made of a collection of simple substances, or elements.

A 16th-century European woodcut shows how ideas about gender, medicine, personality, as well as natural substances, were all combined in the four-element Universe.

Above all others, it was Aristotle, a 4th-century BCE Greek philosopher, who set the trajectory of physics into the modern era. He was a pupil of Plato, and by the time these two greats of Western thought were debating in the walled Athenian olive grove that was Plato's Akademia, the monist concept that the Universe is a infinite swirl of ice, water, and steam, was no long center stage. Instead, an older idea reaching back to earlier civilizations of Mesopotamia and Egypt and shared with those of India and China had been adopted. In the 5th century, Empedocles had refurbished the idea into a Greek form: Nature was indeed constructed of fundamental substances, he said, but water was but one of four, the others being earth, air, and fire. These materials were *stoikheion*, or "components," but have since become known as elements based on the Latin term for "rudiment."

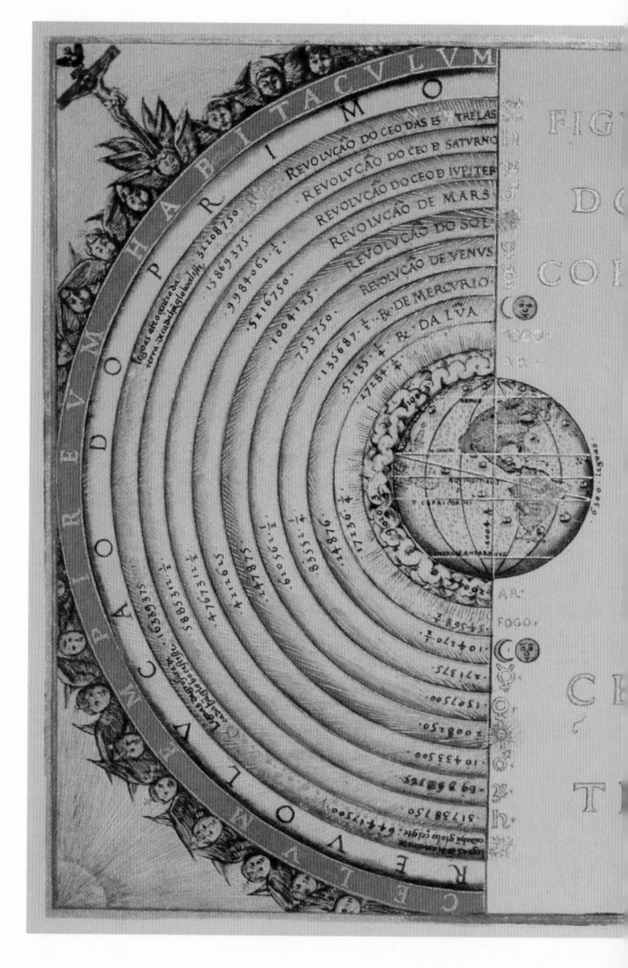

More than a decade after Nicolaus Copernicus published proof that Earth orbits the Sun in 1543, this Portuguese map of the Universe was still based on Aristotle's spheres within spheres with Earth firmly at the center.

The nature of everything

The idea was that everything was made from a mixture of two or more of these elements—and four was the minimum number required to explain the observed nature of things. Damp was evidence of water, heat came from

THE FIFTH ELEMENT

A regular polyhedron is a 3-D object in which all the sides, faces, and angles are equal. The obvious example is a cube, but there are four others: The tetrahedron, octahedron, dodecahedron, and icosahedron. These are better known as the Platonic solids, partly because Plato felt sure that such perfect (and few) geometric forms must be linked to the fabric of the Universe. The first four, therefore, were the shape of the elements, he suggested. But what of the fifth, 20-sided icosahedron? Was this aether (or ether), an all-pervading extra element that filled the space between the first four? The idea of a fifth element, or quintessence (*quintus* is "five" in Latin), proved hard to give up. It was still being proposed in theories into the 20th century.

A detail of the School of Athens painted by Raphael in 1511 for the Pope's Apostolic Palace shows Plato (left) and Aristotle (right) as central figures surrounded by a multitude of great thinkers—and in the main not Christian. Plato points upward because he believes reality is based on imperceptible "forms," while Aristotle gestures ahead—for him it is all about the here and now of tangible objects.

fire, soft objects were filled with air, hard ones made of earth. The conceit was also extended to medicine, and we still nod to those ideas today: A dominance of air-rich blood makes people sanguine airheads; too much water (drools of spittle) makes them phlegmatic wets; yellow bile induces a fiery disposition; while the cold earth in black bile brings forth a parched melancholia. In prescientific medicine good health was achieved by harmonizing these four "humors."

CHINESE CHANGES

The idea of fundamental elements is not a Western one. Classical Chinese thought had five elements: Earth, fire, wood, metal, and water—air was not included. The term "elements" is perhaps misleading here. Certainly, the five were fundamental but in the Chinese world view, they represented stages in an unending cycle of growth, death, and renewal: Wood fed fire; fire created earth (the ash); earth held metal (the ore); metal carried water (impermeable vessels); and water stimulated wood to grow.

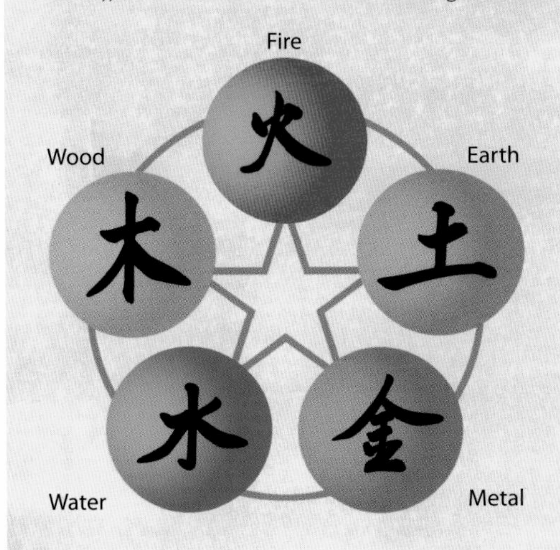

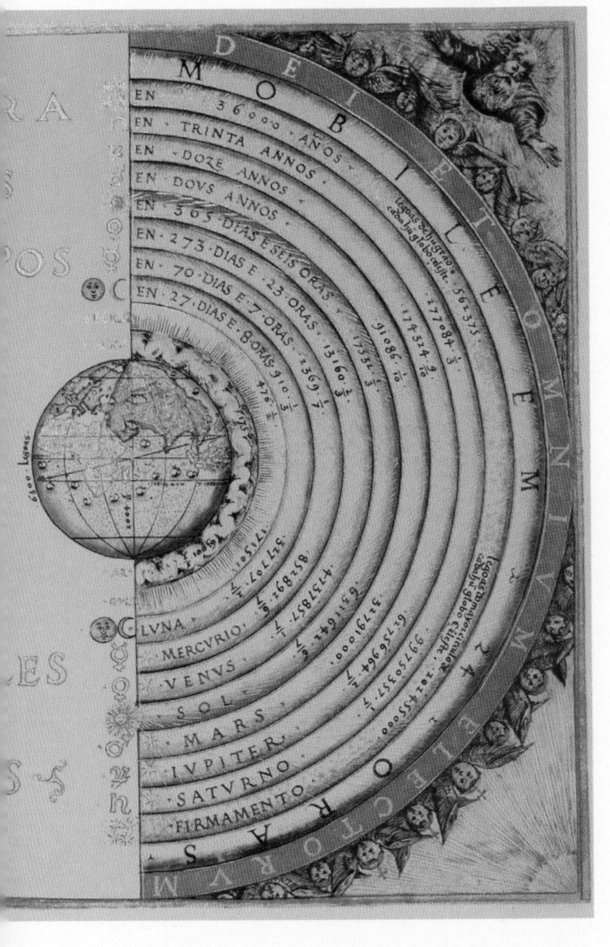

Layered Universe

Empedocles said that elements combined due to the power of love, while strife drove them apart. This eternal battle for harmony is what drove the changes in the Universe. Plato believed that the material world was an illusion of the senses; the elements were in fact perfect, imperturbable forms. Enter Aristotle, who, when viewed on a somewhat superficial level, combined both these ideas.

For him change was the result of elements seeking purity. This was evidenced by the layering of our planet—which he then extrapolated to the entire Universe. The lower layer was earth, forming the rock beneath our feet. Next came water, which formed into oceans, then air, and finally a ring of fire around the planet. Rain was the result of water finding its place in the system and lava was a mixture of air, water, and fire escaping from earth. The four elements of the mortal world reached as far as the Moon. Beyond lay, in ever increasing concentric shells, the Sun, planets, and stars—all made of in ether, a heavenly fifth element. So plausible was Aristotle's Universe it was barely questioned for 19 centuries.

5 Eureka! The Archimedes Principle

ARCHIMEDES CHANGED THE WORLD IN MORE WAYS THAN ANY OTHER ANCIENT SCIENTIST. And his most famous and greatest achievement was made while he was taking a bath.

A playful woodcut from 1547 sums up the Eureka story. Archimedes prepares to take a bath with the washhouse floor littered with metal weights and a rather precious crown. Note the concentric butts. The outer one would collect water displaced by objects dunked in the inner one.

How do solid objects float? This question appears high up on the list when we enquire about the world around us. However, its answer did not come directly from a drive to understand nature—instead, it was the result of a case of suspected fraud. The climax of the story famously took place in a public bath in Syracuse, a Greek colony on the east coast of Sicily. It was here that Archimedes, the ancient world's greatest innovator, had the world's very first eureka moment. *Eureka* means "I have found it!," but what had Archimedes been looking for?

The story was best told in 15 BCE, 150 years or so after it happened, by Vitruvius, a Roman engineer and great admirer of Archimedes. The king of Syracuse, Hiero II, had commissioned a golden crown to be given as an offering to the gods. However, he suspected his purchase was less than pure gold. He consulted Archimedes, undoubtedly the cleverest man in the city, about the problem. Archimedes could not melt down or cut open the sacred crown, so he looked for an alternative test.

CALCULATING PI

The number pi, or π in its Greek letter form, is the ratio of the radius (the distance from the center to the edge) and the circumference (or perimeter) of a circle. Measuring this proportion is difficult; the two lengths never match up. Babylonians made do with an estimate of 3.125 (3 and one eighth) while in Egypt pi equaled 256/81 (3.16). Archimedes turned the Greeks' most powerful tool—geometry—to the task. He imagined a circle to be a polygon with innumerable tiny sides. So by adding more sides to simple polygons and calculating their perimeters, he could edge steadily closer to an accurate length of a circumference and thus calculate pi. He began with a hexagon and then doubled the sides four times to finish with 96-sided polygons inside and out of the circle. That gave him a value for π that lay somewhere between 3.140845 and 3.142857, a result not bettered for 500 years.

Converting a circle into polygons with ever more and more sides edges you closer to a true value for a circumference—but you never get there.

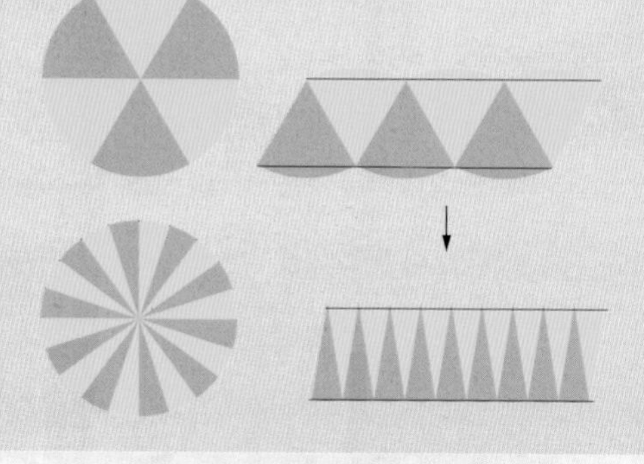

Bath time

Archimedes needed to calculate the crown's density—how much matter was contained in its volume. Comparing this to the density of purest gold would give Hiero his answer. The legend says that as Archimedes sank into his bath one day, he saw water being displaced by his body—and that got him thinking. Immersing the crown in water would displace its exact volume in water. Dividing the crown's volume by its mass would give its density. Eureka!

The displacement method showed that wily old Hiero had indeed been ripped off. The crown's density indicated that it was actually made from an alloy of gold and silver.

How things float

Archimedes did not leave the subject there. In around 250 BCE, he wrote *On Floating Bodies,* stating: "Any floating object displaces its own weight of fluid." A fuller version became what is now known as the Archimedes principle: "The force of buoyancy on an object is equal to the weight of the water it displaces." If the weight of displaced water equals the weight of the object, then it will float under the water. (The object has the same density as water.) If the weight of the object is greater than the weight of the water it displaces, then the object will sink—the buoyancy force is an inadequate counter to the force of gravity (otherwise known as its weight). However, if the weight of the object is less than the weight of the water that it displaces, then it will float on top of the water. Through a series of deductions, others would find that Archimedes had discovered that motion—or the lack of it in the case of flotation—could be the result of opposing forces, a very powerful idea indeed.

ARCHIMEDES' SCREW

This is one innovative device that was not invented by Archimedes. It was probably developed first in Egypt and used to irrigate the Hanging Gardens of Babylon before being named for the Greek after it reached Europe. An Archimedes' screw appears to defy gravity, making water flow uphill. However, it is the combination of two simple machines—a ramp wrapped around a wheel. Turning the wheel raises the water bit by bit up the twisted ramp, and as if by magic it floods out of the top.

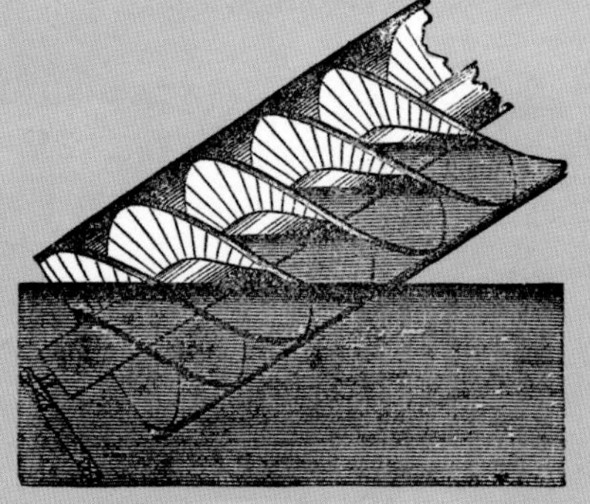

A "golden" crown

Solid gold

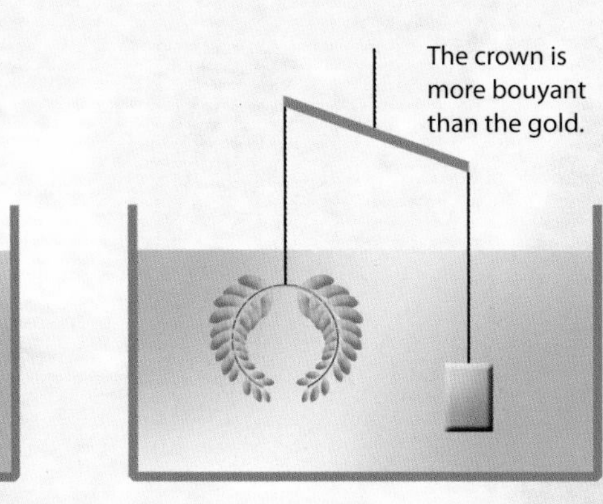

A simple diagram shows the original problem solved by Archimedes' eureka moment. If the crown was pure gold it will have the same buoyancy as a piece of gold of the same weight. However, if the crown is more buoyant—displaces less water than the gold—then it is made from cheaper, less dense metals.

6 Making Machinery

PHYSICS IS NOT JUST THE FOUNDATION OF ALL SCIENCE, BUT IT IS WHERE TECHNOLOGY STARTS TOO. One of the first appliances of science was used in that other great Greek cultural legacy—the theater.

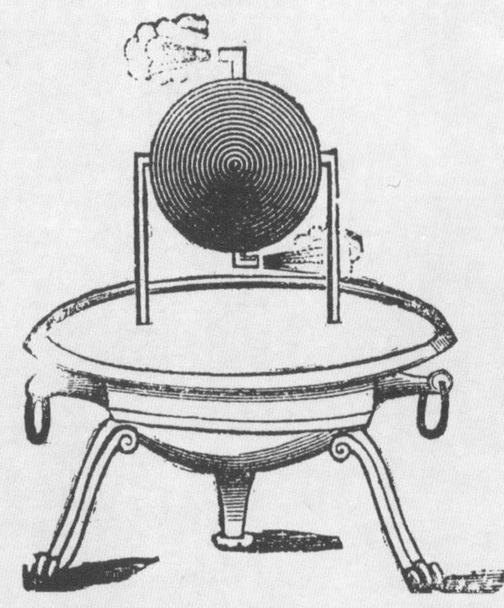

Hero's "aeolipile" was a brass sphere fitted with two angled nozzles on opposite sides. Steam rising from a boiler beneath escapes from the nozzles, causing the sphere to spin. It makes use of the action and reaction of forces, a universal law of motion codified by Newton 15 centuries later, and unknown to Hero.

Greek inventors were the original mechanics. The word *machine* comes from "mechane," a device used in performances to lift actors high above the stage, generally in the guise of some god or other. No designs of a mechane survive but it is assumed they used a combination of crane-sized levers and simple pulleys.

Dating from at least the 4th century BCE, they predate Archimedes (3rd century BCE), who is said to have invented the compound pulley and famously quipped of levers: "Give me a lever long enough and I will lift the world." Archimedes put his discoveries into service in the defense of his city from Roman attack in the Punic Wars. His "ship-shaking" invention made use of them all. It hooked onto a ship and used the mechanical advantage of pulley and lever to lift, or at least tilt, one end of the enemy vessel. That let water flood in the other end—the result being a change in buoyancy, which took the ship to the bottom!

GETTING MECHANIZED

Ancient Greeks loved mechanical toys called "automata," powered by turning a handle. Hero of Alexandria built a ten-minute automata show, which may have inspired this idea for an entirely mechanical blacksmiths, where waterpowered automata hammer metal.

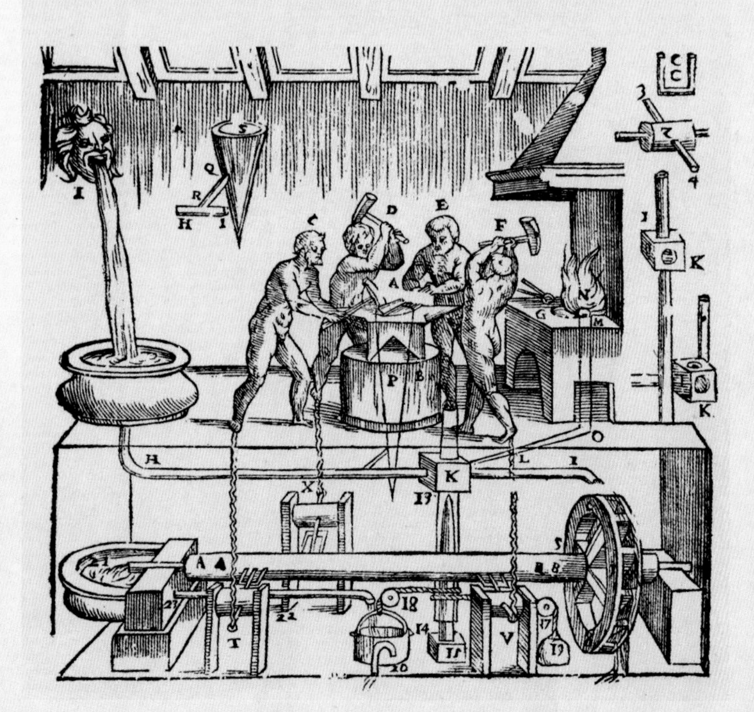

Force a reaction

The natural successor of Archimedes is Hero of Alexandria, who lived in the 1st century CE. One notable device of his was a "windwheel" used to harness the power of wind to play an organ. Another was coin-operated vending machines for temples—the weight of the coin raised a lever, releasing a small gush of holy water. But Hero is best remembered for inventing the steam engine, or at least an engine that used steam. The device was forced into a spin by jets of steam. That makes it a reaction engine, working like the nozzle of a rocket. Hero found little use for his invention other than as a curiosity. And a curiosity about how it moved would one day lead to the scientific field of mechanics—the physics of force and motion.

7 Seeing Beams of Light

"SEEING IS BELIEVING," THEY SAY, BUT HOW CAN WE TRUST OUR EYES? We have a Arab scholar imprisoned in 11th-century Cairo to thank for it.

The title-page illustration of a 13th-century reworking of al-Haytham's Book of Optics *by Polish author Witelo shows Archimedes' "heat ray" in action. Legend has it the Greek used bronze parabolic mirrors to focus the Sun's rays on Roman ships, setting them alight. Alhazen's book discusses the optics of these curved mirrors in detail.*

Ibn al-Haytham—sometimes known by the Latinized name Alhazan—did what many in the 21st century still fail to: He made the distinction between astrology and astronomy. The former lacked evidence, while the second relied wholly on the movements of lights in the sky, and it was the study of light that would make his name. His no-nonsense approach led to the Iraq-born scholar being summoned to Cairo by the Fatamid caliph in 1011. His task was simple: Build a dam across the Nile. Simple to say, at least, but al-Haytham soon realized it was impossible to do, and decided to feign insanity to get out of the job without incurring the wrath of the caliph—the most powerful man on Earth at the time.

Straight lines of light

Al-Haytham was kept under house arrest for the next decade, and it was at this time that he did the work that was published as *Kitab al-Manazir* (*Book of Optics*). In it al-Haytham proved that light travels in straight lines using one of the first scientific experiments. He observed a light through a hollow tube. He then blocked the end, and the light was no longer visible. Obvious as it may be, but this was the first empirical proof that light could take only one straight path to the eye. It allowed al-Haytham to describe light rays using geometry, an early example of math used in science.

Until this time, vision was explained as either the eye sending out straight rays that were reflected from the surroundings, or as objects transmitting fully formed images that somehow entered the eye. Al-Haytham combined both into a correct description: Light travels from objects and enters the eye as straight rays, forming an image on the back of the eyeball.

THE CAMERA OBSCURA

Al-Haytham provides the first description of a *camera obscura*, a room-sized pinhole camera. He saw the same effect in natural settings, such as the suns projected through gaps in leaves on to dappled forest floors. He also noted that tiny pinholes produce the best images.

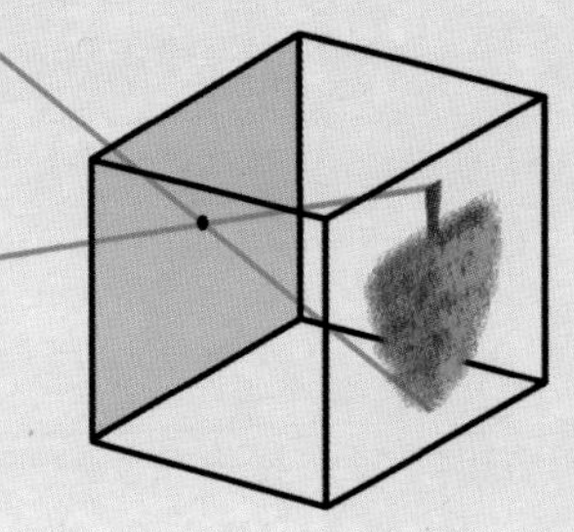

8 The Mechanics

Physics is based on an intuition that the Universe works according to rules. In the 12th century, Islamic scholars reasoned that the rules that made things move also made them stay still.

For 14 centuries, few scholars questioned Aristotle's theory of motion. This detailed how a moving object was cutting through the air at its leading edge, while simultaneously being pushed along by the air that rushed in behind it. In other words, motion was the result of a constant force. Al-Biruni, a Persian scholar, questioned this theory when noting that a body moving in a non-uniform way—changing direction or speed—must undergo a period of acceleration. Over in Baghdad, Abu'l-Barakat added to this idea by saying that motion was the result of a single short push, while a constantly applied force resulted in acceleration. A century later, the Persian Al-Khazin investigated weights and balances. He built on the work of his predecessors to unify the ideas of dynamics (unbalanced forces creating motion) and statics (balanced forces) into the single field of mechanics. All were force in action.

An Arabic manuscript from around the 16th century shows the application of mechanics. It contains a diagram of a machine for raising water that suggests a wheel turned by a flow of water is connected through gears to a series of buckets on a belt.

9 Force and Inertia

Ibn Rushd, known as Averroes in Europe, was an Islamic polymath from 12th-century Spain. Among many contributions, he proposed an early version of inertia, now central to the physics of motion.

Averroes wrote three commentaries on the physics of Aristotle. So influential were they that later European scholars referred to Averroes simply as the Commentator. In the commentaries, Averroes defined force as something that changed the motion of a body. He also made the link between the magnitude of force and the rate of change of motion. Crucially, he proposed that bodies have an inherent resistance to changes in motion, separate to the drag of air or effects of gravity. Averroes based this idea, which we now understand as "inertia," on the fifth element, ether. According to Aristotle, ether only exists beyond the Moon, and so Averroes thought only heavenly bodies had inertia—that is why they do not move at infinite speeds. However, inertia would eventually be regarded as a universal feature of all matter.

A portrait of Averroes based on his appearance in Raphael's fresco, School of Athens, *in the Vatican. He is surely one of the few Islamic figures to be depicted in the Pope's residence, which reflects just how significant his work was to the advancement of knowledge in a wide number of fields.*

10 Artificial Rainbows

ALL EXPLANATIONS OF RAINBOWS, A VERY COMMON OPTICAL PHENOMENON, were failures, until a German monk decided to make his own raindrops from glass.

Meteorology, the science of weather, is a complex field. It is little wonder that the ancients put shooting stars and comets into the same set of phenomena as hail, lightning, and fog. When it came to rainbows even al-Haytham got it wrong. His suggestion was that a rainbow was a partial image of the Sun reflected off a curved mirror formed by the water droplets in a cloud. Only in the year 1300, did Theodoric of Freiberg give an alternative. He recreated raindrops as glass globes filled with water. He saw that a light beam was redirected, or refracted, onto the back of the "raindrop," where it was reflected out again. A second refraction as it left the droplet split the white light into its constituent colors. The rainbow is the cumulative effect of all these refractions and reflections.

Rainbows hold a special place in all cultures, with colorful interpretations such as a heavenly water snake, necklace of a goddess, or a bridge between heaven and Earth. If the Sun comes out while it is still raining, turn your back to it to see the rainbow—optics in action.

11 Ockham's Razor

THERE IS A SKETCH OF THE ENGLISH PHILOSOPHER WILLIAM OF OCKHAM IN ONE OF HIS BOOKS WRITTEN IN 1323. It shows him to be clean shaven, an example of Ockham's Razor in action. Or are we being too simplistic?

Ockham's Razor has nothing to do with toiletries. It is a rule that guides scientists to develop sound explanations, a mental blade that cuts away superfluous assumptions. It carries Ockham's name more out of accident than achievement, although the English friar was one of many to comment on it in his writings. Also known by the Latin term, *lex parsimoniae*, meaning the "law of parsimony," the best formulation of Ockham's Razor belongs to Thomas Aquinas who lived a generation before William: "It is superfluous to suppose that what can be accounted for by a few principles has been produced by many." In other words, start with the simplest theory and only add complexity if needed—

12 Adding Impetus

PARIS HAD ONE OF THE FIRST UNIVERSITIES IN THE WORLD, ALREADY A CENTURY OLD BY THE TIME WILLIAM OF OCKHAM went there to teach in the 1320s. And one of his students would add some momentum to physics in more ways than one.

Buridan said that impetus would keep a body in motion forever unless a resisting force acted to diminish it.

Jean Buridan studied under Ockham before joining his mentor on the academic staff in Paris. His personal life appears to have been eventful, and his work also caused a stir. Buridan used the breakthroughs of Islamic mechanics to make a blatant break with Aristotle's teachings, a controversial act in Europe. He said that when a body is set in motion by a force, it keeps moving due to its "impetus." The impetus stays with the body, and motion only ends when another opposing force is big enough to take it away. This is on the way to the modern idea of momentum, which relates the quantity of a body's motion.

13 Theory of Tides

TODAY, WE KNOW THAT THE TIDES ARE CAUSED BY THE GRAVITY OF THE MOON AND SUN PULLING ON THE OCEANS. But in medieval times gravity was a force that just pulled things down—so what was moving water in and out?

Seen at a local level, tides make water appear to move in and out from shore. From a global perspective the ocean water is pulled into an ovoid, which means the height of water's surface can vary by several feet.

The link between the location of the Moon and the ebb and flow of the tides was made in ancient times. However, the rhythmic cycles of the Moon alone did not correlate with the spring and neap tides. In 1523, Italian Federico Grisogono outlined how the pull of the Moon would turn a sphere of water into an elongated ovoid. A secondary weaker pull from the Sun could alter that elongation: Neap tides arose when the pulls of the Moon and Sun were perpendicular; spring tides formed when they aligned. The question was what caused the pull? Was it gravity or perhaps magnetism?

14 Understanding Magnets

IN 1600, AN ENGLISH DOCTOR SHOWED THAT EARTH EXERTED FORCES OTHER THAN GRAVITY, when he showed our planet is a giant magnet.

A Greek philosopher Theophrastus wrote about *magnítis líthos*—the "stones of Magnesia"—in the 4th century BCE. Magnesia is a region of Greece, and the stones in question were lodestones, iron-rich rocks that are naturally magnetic.

Magnets were put to use as compasses in China in the early centuries of the first millennium CE, although initially for rituals and divination. By the 11th century, the Chinese were using compasses for navigation, a technology that spread across Asia and Europe over the next 400 years.

DE MAGNETE, LIB. IIII. 155

CAP. II.

Quòd variatio ab inæqualitate eminentium telluris partium efficiatur.

DEmonſtratur hoc ipſum manifeſtè per terrellam, hoc modo : ſit lapis rotundus aliquâ parte imperfectior, & marcore labefactatus (talem habuimus parte quâdam carioſâ, ad ſimilitudinem maris Atlantici, ſiue Oceani magni) pone fila ferrea longitudinis granorum duorum hordeaceorum ſuper lapidem, vt in ſequente figura. A B, Terrella partibus quibuſdam imperfectior, & virtute in circumferentia, inæqualis:

A

B

Verſoria E,F, non variant; ſed directè polum A reſpiciunt : poſita ſunt enim in medio firmæ & valentis partis terrellæ, longiùs ab imperfectâ : ſuperficies punctis & lineis tranſuerſis inſignita, imbecillior eſt. O (verſorium) etiam non variat (quià in medio imperfectæ partis) ſed in polum dirigitur, non aliter atq; iuxta occidentales Azores

William Gilbert's discoveries were published in the 1600 treatise, De Magnete *(On the Magnet).*

On the Magnet

William Gilbert, the physician to the English queen Elizabeth I, had a sideline in studying the mysterious forces of magnetism and electricity. (He even gave the latter phenomenon its name, based on *elektron*, the Greek word for "amber," a material that becomes charged with static electricity when rubbed.)

In 1600, Gilbert revealed that compass needles point north because our entire planet is a magnet. Before that it had been suggested that compasses were attracted to the Pole Star or a mysterious iron island in the far north. Gilbert proved that a compass needle and Earth work according to the same laws of attraction and repulsion that govern the interaction of any two magnets. He did this with a "terrella," a model world carved from a lodestone. A compass placed on the surface of the terrella behaved just as it would when used on Earth itself for navigation. Magnetism and electricity were forces of nature like gravity. Were they linked in some way? And were there any others left undiscovered?

POLE EXPLORER

The compass is a Chinese invention, and its use in navigation is down to Shen Kuo, who recorded his achievements in the wonderfully exotic-sounding *Dream Pool Essays* of 1088 CE. In it Shen makes the first explicit reference to a compass needle finding north, once allowed to move freely. (In modern terms the north pole of a magnet is attracted to Earth's North Pole.) However, Shen also found that compasses point a few degrees west of true north. Earth's magnetic field is slightly wonky.

15 Law of Refraction

LIGHT BENDS, OR REFRACTS, AS IT PASSES FROM ONE TRANSPARENT MEDIUM TO ANOTHER. The apparent kink in a drinking straw in a soda is proof enough, although more ancient evidence came from light curving through bowls of water. Was there some kind of pattern to be seen?

INTERNAL REFLECTION

If a light beam arrives at an interface between media at an angle above a critical value, it will not refract through it, but reflect back. This phenomenon, known as "total internal reflection," is what makes cut gems sparkle and the ocean glitter.

Diamond has a refractive index of 2.4, one of the highest of any naturally occurring mineral. The cuts of diamond shape the stone's surfaces so that light traveling inside the gem is refracted and reflected out the top, giving it that expensive-looking sparkle.

We have Hero of Alexandria to thank for the law of reflection, which states that when a beam of light strikes a smooth reflective surface, it will reflect back at exactly the same angle—albeit on the other side of an imaginary vertical dividing line. This is why a reflection of your face forms a mirror image, where the left side appears on the right, and vica versa.

Scholars searched for a similar law that governed refraction, a phenomenon in which a beam of light deviates as it crosses the boundary between two media—from air to glass, for example. Al-Haytham and others had seen that light crossing a curved boundary, such as a glass bowl, was not all refracted by a uniform angle. A larger initial angle resulted in a larger deviation. This was why glass bowls acted as lenses, focusing once parallel rays of light into a single point.

Going Dutch

The Persian Ibn Sahl had, in 984, figured out how to calculate angles of refraction using ratios already, but we still know it as "Snell's law" after the Dutchman Willebrord Snell, who rediscovered the law in 1621. The law requires that all media have a refractive index, which is really the ratio of the speed of light through that medium compared with through a vacuum (which has an index of 1). In the end, refraction is the result of the beam changing speed as it moves across the boundary between media. Light hitting at an angle of 0° (to the vertical, that is) never refracts—it just carries straight on, albeit at a different speed. Light arriving at an angle changes course because one section crosses the boundary and changes speed before the other. Light moving into a medium with a higher index will refract to a smaller angle; a lower index produces a larger angle.

$$n_1 \sin\theta_1 = n_2 \sin\theta_2$$

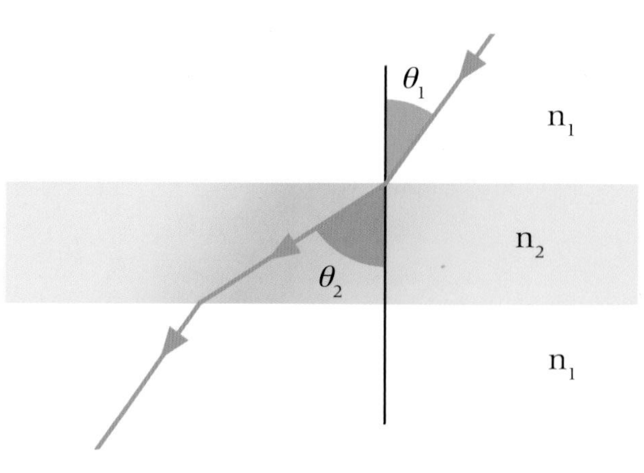

Snell's law makes light work of the angle of refraction: n_1 and n_2 are the refractive indices of the media, θ_1 is the initial, or incident, angle, and θ_2 is the refractive angle. As the light leaves the second medium, the calculation is effectively reversed.

16 Galileo: The Fall Guy

LEGEND HAS IT THAT GALILEO GALILEI CLIMBED THE LEANING TOWER OF PISA TO EXPERIMENT ON THE WAY OBJECTS FELL FROM THE TOP. He almost certainly never did this, but the great Italian scientist was the first to use quantified data to reveal some universal truth about what was happening in free fall.

After centuries of scholars looking for wriggle room within the teachings of Aristotle to explain how bodies really moved, Galileo swept away all those previous assumptions on motion and started again. The result is less well known than his later astronomical observations which showed that not everything orbited Earth, but they are just as important to the progress of our understanding of the Universe.

Galileo was only interested in evidence that he could gather himself. The Leaning Tower "experiment" was probably a story Galileo used to debunk Aristotle's theory of motion. This stated that heavier objects fall faster than lighter ones. His other experiments with balls and ramps (this time certainly real), showed that a falling (or rolling) body was subject to a constant force and thus was accelerating—speeding up at a constant rate. This acceleration (*a*) resulted in the distance (*d*) the body traveled, being proportional to the square of the time (*t*) it spent in motion. Unlike his predecessors, who suspected as much but failed to prove it, Galileo was able to express this law of fall in unequivocal mathematical terms: $d = at^2$. If modern physics begins anywhere, it begins here.

Galileo takes center stage in this 1840s' depiction of a demonstration in Pisa of the law of fall. His ideas are opposed by both a noble patron (right) and assorted clergymen and philosophers, but Galileo's ramp experiment wins the day.

17 Applying Pressure

"NATURE ABHORS A VACUUM" IS A CONCEPT DATING BACK TO CLASSICAL GREECE. Back then it meant that a vacuum was simply impossible, but investigations into the forces acting on liquids and gases—not least by pumps—would lead to a new understanding.

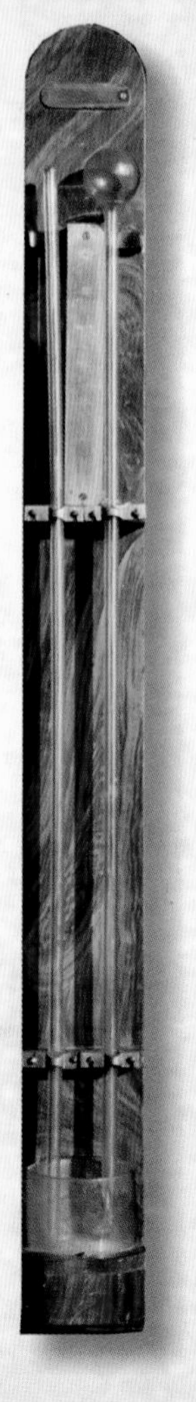

A reconstruction of one of Evangelista Torricelli's barometers. The left side of the U-shaped tube is open to the air, which pushes down on the mercury, forcing it up on the right side. The level that the mercury reaches gives a reading of the air pressure.

Like so many scientific stories from the 17th century, this one begins with Galileo. At the height of his powers in 1630, Galileo received a letter from fellow Italian Giovanni Battista Baliani. In it Baliani relates how it was proving impossible to pump water through a siphon over a particularly large hill. The reasoning at the time was that a pump would draw water through a siphon by creating at least the possibility of a vacuum. The water would rush in to fill the space, moving through the pipe as it did so. Galileo suggested that even the power of a vacuum had a limit and it was simply not strong enough to haul water over what in modern terms was equivalent to a six-story building.

Three years later Galileo was tried for heresy by the Inquisition in Rome. He had made a career of refuting Aristotle's world view, a view that put Earth at the center of a perfect, unchanging Universe and which had been adopted by the Catholic Church. Galileo was sentenced to house arrest, a punishment that lasted until his death in 1642. A few months before he died, Galileo (now blind) invited the mathematician Evangelista Torricelli to be his aide in recording his final works. Torricelli then filled Galileo's role as Tuscany's chief scientist.

Building a barometer

One of the first problems Torricelli was consulted about was why there was a limit to the height even the largest suction pumps could raise water (around 10 m or 33 feet in today's terms). Torricelli chose to study the problem in miniature, scaled down ten times. He closed one end of a glass tube and filled it with mercury—a liquid that is 14 times as dense as water. He then placed the open end of the tube into a bowl of mercury (pictured right). The mercury in the tube always fell to 76 cm (30 inch). It appeared that a column of mercury had a maximum height too, and it was approximately 14 times smaller than that of a water column. This was enough evidence for Torricelli to turn the theory of pumps and vacuums on its head.

He found that a liquid was not raised by the pull of a vacuum but pushed by the weight of the air. A column reached its maximum height when its weight was balanced with that of the air. This earned Torricelli the rights to the invention of the barometer—a device for measuring air pressure—although others had tinkered with them before. Pressure is a measure of force acting over a surface area.

Blaise Pascal repeated his pressure experiment in person at the top of the 50-m (164-foot) tower of Saint-Jacques-de-la-Boucherie, a Paris church. In honor of the Frenchman's work on pressure, the unit of pressure is named the pascal (Pa).

Vacuum power

What was the space left at the top of the "Torricellian tube"? What else could it be but a vacuum? With Torricelli sent to an early grave by typhoid in 1647, the following year Blaise Pascal took over the investigation. Giving directions from decidedly flat Paris, he had his brother-in-law Florin Périer perform a now legendary experiment in the Massif Central mountains. Perier set up two mercury barometers outside a monastery in Clermont-Ferrand and tasked a monk to record the level of one throughout the day. (It did not change.) Perier then hauled the other barometer to the top of Puy de Dôme—an extinct volcano nearby standing 1,460 m (4,790 feet) tall. Perier made meticulous measurements on the way up and found that the level of the mercury dropped as he climbed. This was as Pascal had predicted. The air pressure dropped with altitude as there was less air pushing down from above. Moreover, the space above the mercury grew despite being sealed from the outside world. Only nothing could grow by adding nothing. Here was a vacuum, the empty space so abhorred by the followers of Aristotle.

Von Guericke recreated his most theatrical of experiments for Frederick William I, Duke of Prussia, near Berlin in 1663, as depicted in this engraving in von Guericke's book, Experimenta Nova.

Getting pumped

In 1650, the action moved again to Magdeburg, Germany. Here Otto von Guericke had invented a vacuum pump with innovative one-way valves. He famously used it to suck the air out of two copper hemispheres that had been fitted together. The story goes that eight-horse teams harnessed to both hemispheres could not pull them apart. Once air was allowed back into the metal sphere, the two halves were easy to separate. Here was yet more proof that a vacuum exerts no force. Instead, the pressure of the air was holding the hemispheres together. The next question was: What was it in the air that did the pushing?

18 Pendulums

LEGEND HAS IT THAT GALILEO DISCOVERED THE REGULAR MOTION OF PENDULUMS while watching a lamp set swinging in Pisa Cathedral. The natural rhythm of pendulums revolutionized timekeeping and also gave further insight into the principles of motion.

Christiaan Huygens patented pendulum clocks in 1656, and had them made under contract.

GALILEO'S LAMP

A lamp still hangs in Pisa Cathedral where Galileo was inspired to launch his research into the motion of pendulums. The story goes that Galileo used his pulse to time each swing during services. However, tourists beware. The current lamp dates from 1586—four years after Galileo's epiphany.

Whatever the truth behind this charming story, it is certain that Galileo made a thorough study of the motion of pendulums early in his career. By around 1602, he had found that the period of a simple pendulum is proportional to the square root of its length. (The period is the time it takes for the pendulum to complete a single oscillation, or swing in both directions. It is best imagined as the time needed to move from the extreme left, through the central point to the extreme right, and all the way back again.) Changing the mass of the pendulum bob has no effect on the period—a heavier bob oscillates with the same frequency as a lighter one. Crucially, Galileo also discovered that for small swings, the period was also independent of amplitude—width of the swing. That meant all pendulums of the same length, set swinging anywhere in the world, would always have the same period. This property of "isochronism" (meaning equal time) made it possible to use pendulums for timekeeping. All that was needed was a pendulum with a period of one second. Galileo had changed the world again.

SIMPLE HARMONIC MOTION

The motion of a pendulum is an oscillation. Weighted springs left to bounce up and down oscillate in the same way. In ideal terms, when none of the energy is lost through friction, these oscillations can be modeled as simple harmonic motion. This is where the velocity (*v*) of the weight, its acceleration (*a*), and its displacement (*s*) from the central start point are all in perfect proportion. In reality, friction will always dampen the oscillation, bringing it to a halt sooner or later.

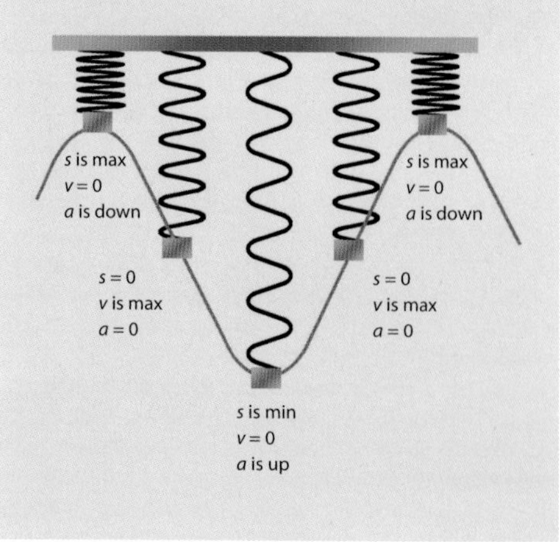

Tick and tock

Knowing the time was most crucial for medieval clergy who framed their activities around prayers at certain points of the day. Mechanical clocks had been around since the 14th century—Milan had one in 1335, for example. They operated by means of a weight falling at a controlled rate of descent. They were far from accurate and had to be frequently reset using astronomical observations.

Galileo left an unrealized design for a pendulum clock on his death in 1642—but it was Dutch scientist Christiaan Huygens whose 1656 design was in tune with the manufacturing abilities of the age. Huygens' clocks were powered by the pull of a weight as in earlier designs, but used the swing of a pendulum to release it. This allowed existing church clocks to be retrofitted with a more accurate device. Fine tuning was achieved by moving the bob up or down to adjust the period to exactly one second, and even these simple pendulum clocks lost only seconds a day.

Pendulums are an example of harmonic oscillators. The swing is produced by a restoring force, so-called because it is always pulling toward a central point, and the magnitude of that force is always proportional to the distance from the central point. Next, a young English scientist applied the same idea to the way things stretched.

19 Hooke's Law

Robert Hooke was destined to be in the shadow of greater figures, but he has many scientific discoveries to his name. Not least is the law that relates how things stretch and compress.

Robert Hooke was on the stage during many great moments of scientific discovery—but the spotlight was frequently on others. He was regularly in dispute with his colleagues over who had discovered and invented various things. It was to one such a complaint by Hooke that Isaac Newton replied, "If I have seen further it is by standing on the shoulders of giants." Hooke was probably a hunchback, and far from giant.

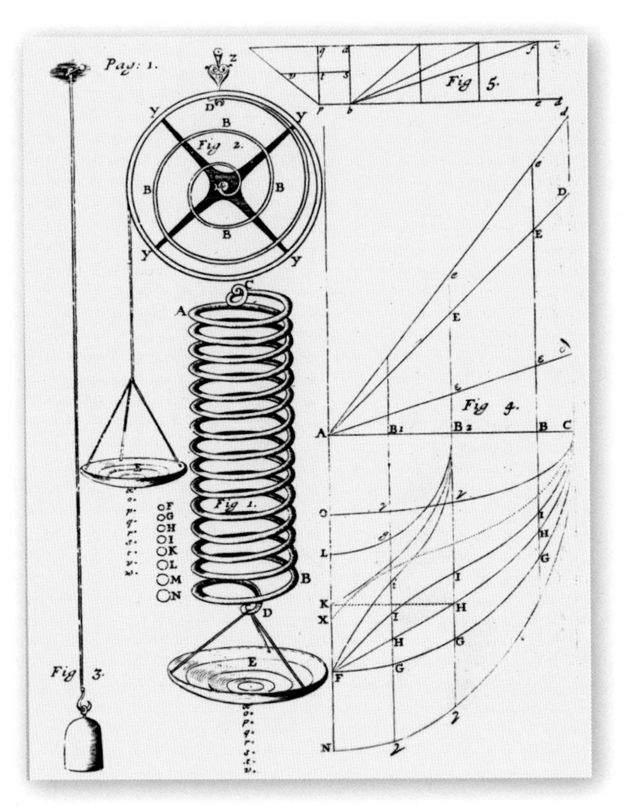

Stretching reality

However, as its name suggests, Hooke's law is all his, even though he chose to announce it in 1660 as an anagram: *ceiiinosssttuv*. This can be rearranged into *ut tensio, sic vis*, which is Latin for "As the tension, so the force." Put simply, the extension (or compression) of a given material is proportional to the force applied—a "restoring force" needed to return the material to its initial length: Double the force, double the extension. Obviously, some materials are stiffer than others, but the law holds until the material begins to deform and break. Hooke did much of his work with springs (he argued with Huygens about using spring oscillators in clocks), but his law applies to the plucking of a string, earthquake tremors, and the vibrations of an atom.

SCIENTIFIC MONUMENT

The Monument to the Great Fire of London was built by Robert Hooke and Christopher Wren, Britain's most iconic architect. The column, which marks where the fire started in 1666, was designed for much more than that. The steps of the spiral stairwell were used to measure elastic extensions and oscillations, while an opening in the roof made it a zenith telescope.

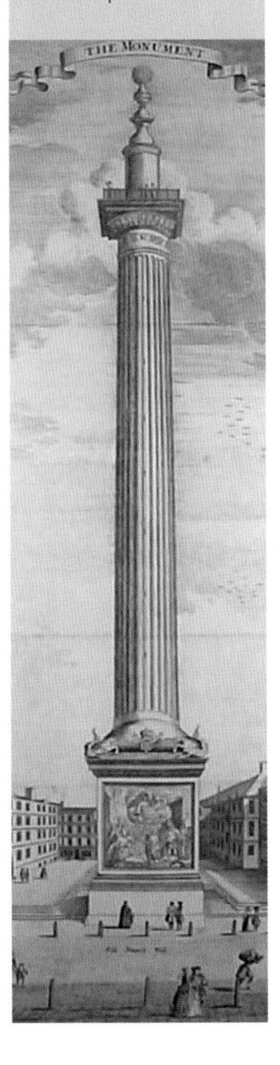

20 Gas Laws

ALTHOUGH HE IS SAID TO BE THE "FATHER OF CHEMISTRY," ROBERT BOYLE'S famous gas law is pure physics as it describes how a gas's pressure relates to its volume. This and two more gas laws provided the first clues as to how matter was constructed on the as-yet inconceivable atomic scale.

In the late 1670s, Robert Boyle was assisted by Frenchman Denis Papin, seen on the right in a laboratory filled with glassware and other apparatus, including the spherical vacuum pump (back right) so central to Boyle's discoveries.

While Pascal studied forces on liquids and Hooke had revealed a feature of solids, Robert Boyle chose to follow von Guericke's path and investigate gases, or "airs," as they were known then. (Remember, even then there were few alternatives being offered to a nature comprised of the four classical elements: Air, earth, fire, and water.) Boyle was also heavily influenced by Francis Bacon, an English courtier who published one of the first descriptions of a scientific method—although others, not least al-Haytham and Galileo, had been working along similar lines already. Boyle began a tradition of skepticism that permeates through science even today. Specifically, he refused to take for granted the claims of alchemists, which were as much based on superstition as fact.

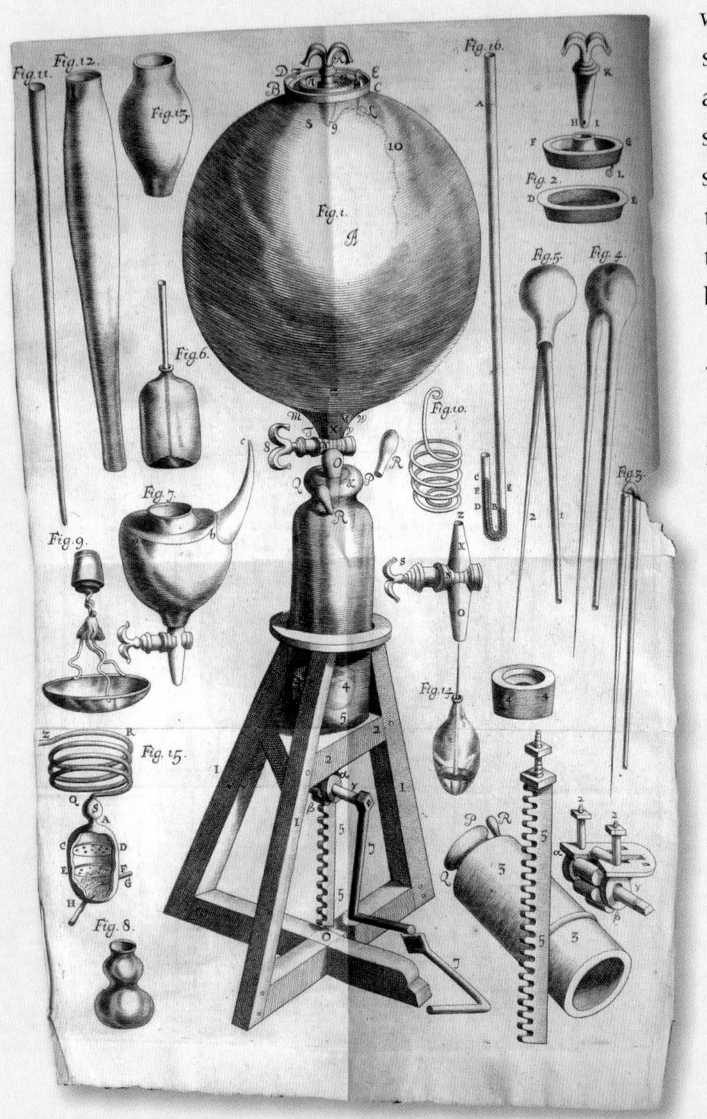

Touching the spring of the air

Coming from a wealthy Irish family, Boyle set himself up as a gentlemen scientist, building a laboratory in his London home. He employed Robert Hooke to retroengineer a "pneumatical engine"—a vacuum pump like von Guericke's.

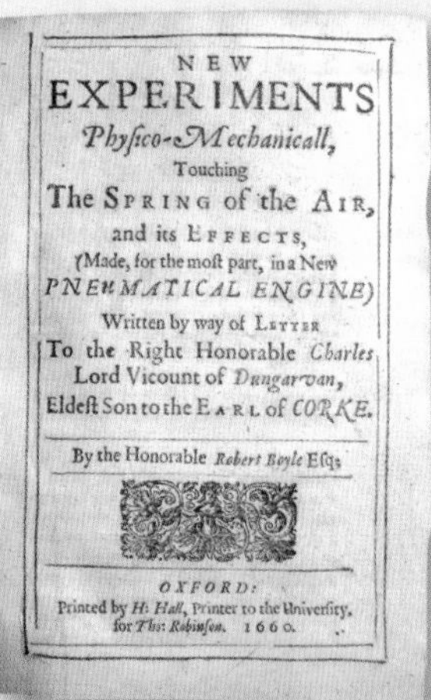

NEW
EXPERIMENTS
Physico-Mechanicall,
Touching
The SPRING of the AIR,
and its EFFECTS,
(Made, for the most part, in a New
PNEUMATICAL ENGINE)
Written by way of LETTER
To the Right Honorable Charles
Lord Vicount of Dungarvan,
Eldest Son to the EARL of CORKE.
By the Honorable Robert Boyle Esq;
OXFORD:
Printed by H: Hall, Printer to the University,
for Tho: Robinson. 1660.

Boyle's first book, published in 1660, was the descriptively titled New Experiments Physico-Mechanical: Touching the Spring of the Air and their Effects and included detailed illustrations of his ground-breaking "pneumatical engine."

Boyle also had the city's best glassmakers create a veritable wealth of unusual glass vessels to aid in his first round of experiments. These were published in 1660 under the title of *New Experiments Physico-Mechanical: Touching the Spring of the Air and their Effects*. The experiments showed that sound could not travel through a vacuum, flames did not burn in the absence of air, and animals and plants could not live without air.

Boyle also investigated the physical nature of "airs." For example, he showed that in a vacuum a feather fell just as fast as a stone. This suggested that although it was invisible, the air contained matter. In 1662, Boyle published what became known as Boyle's law. He had found that the pressure of a vessel of gas increased when it was squeezed into a smaller volume. Expressed in modern terms, gas pressure (P) is inversely proportional to volume (V): $P \propto 1/V$.

Boyle used his empirical law to promote the idea that air was made up of corpuscles, tiny masses that moved in all directions, bouncing off each other and colliding with the walls of their container—creating pressure.

Other laws

More than a century later, two more gas laws were added to Boyle's that tackled temperature (T). Charles's law (1780) says that gas temperature is proportional to volume ($V \propto T$), while Gay-Lussac's law (1802) states that gas pressure is proportional to temperature ($P \propto T$). Together, the three gas laws form the foundations for modern atomic theory. As Boyle had suggested, the behavior of airs—eventually renamed "gas," based on the Greek word *chaos*—was best explained by treating them as collections of independent units. In hindsight, the course to the discovery of atoms appears to have been set but it would take centuries more to confirm.

THREE LAWS EXPLAINED

Gay-Lussac's law states that if the volume is constant, heating a gas will increase the pressure. This is because the gas molecules are moving faster and hitting the sides of the container harder and more often—and exerting a greater force.

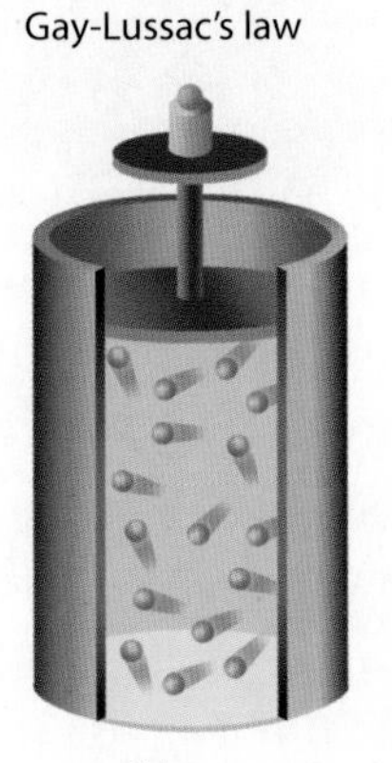

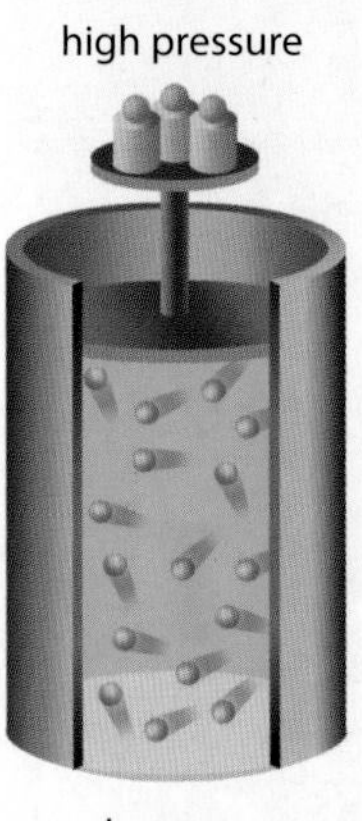

Charles's law states that gases expand as they get hotter. The hotter, faster molecules exert a pressure against the container, making it expand until reaching a new equilibrium.

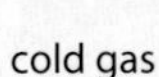
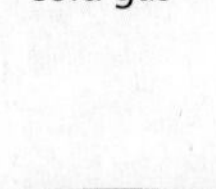

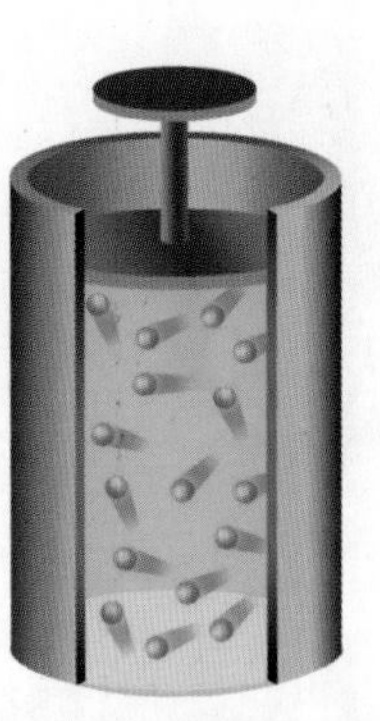

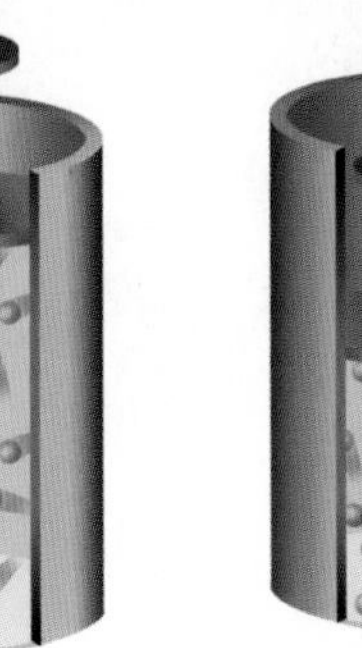

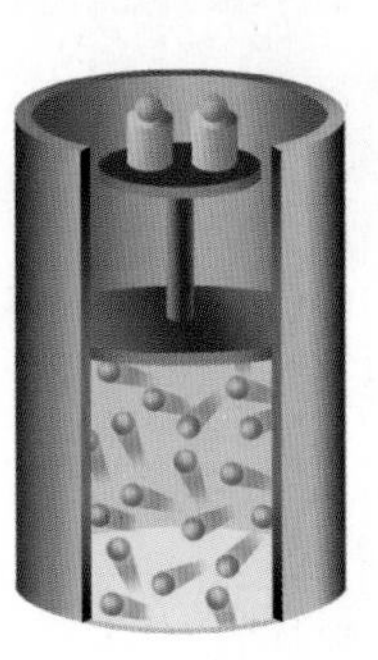

Boyle's law states that the pressure of a gas is inversely proportional to its volume. When you squeeze a gas into a smaller volume, its pressure goes up because its molecules are hitting the inside of the container more frequently.

THE SCIENTIFIC REVOLUTION

21 Newton's Principles

History is written by the victors, and Isaac Newton was one of life's winners. His most important work, published as *Principia* (meaning "principles"), changed the way we understand motion and the force of gravity.

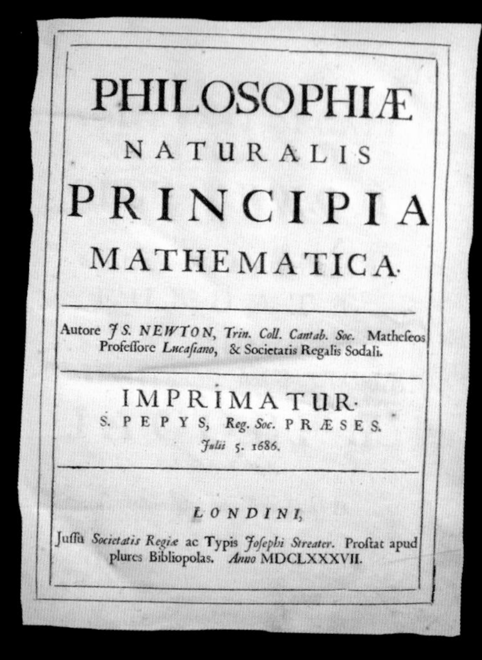

PHILOSOPHIÆ
NATURALIS
PRINCIPIA
MATHEMATICA.

Autore J S. NEWTON, *Trin. Coll. Cantab. Soc.* Matheſeos Profeſſore *Lucaſiano*, & Societatis Regalis Sodali.

IMPRIMATUR.
S. PEPYS, *Reg. Soc.* PRÆSES.
Julii 5. 1686.

LONDINI,
Juſſu *Societatis Regiæ* ac Typis *Joſephi Streater*. Proſtat apud plures Bibliopolas. *Anno* MDCLXXXVII.

The title page of the first edition of Principia.

An engraving of Newton in his middle age by which time he had already recalibrated physics and mathematics. He went on to a career in politics and government.

The legend has it that the idea for gravity flashed into young Isaac's mind in the 1660s, when an apple fell from the tree that he was sitting under. Newton only thought to mention this story in his eighties (and after a heavy dinner) but it seems to make him appear as some kind of seer whose sparkling genius brought about the Scientific Revolution on its own. In his lifetime, Newton was vehement in the defense of his achievements but he also acknowledged the contributions of earlier scholars.

One force needed

In 1609, Johannes Kepler, a German astronomer, found that the orbits of planets were ellipses, not circles, as had been previously assumed. A little while later Galileo extended his work on falling bodies to consider the path of projectiles, and found that they followed a curve called a parabola. It had been known since the Greeks that ellipses and parabolas were related shapes, and Newton's first step was to realize that the force that kept planets in orbit was the same one that pulled projectiles back to the ground. Newton kept this theory to himself for years. In the 1680s, Robert Hooke suggested to his colleague Edmond Halley (he of the comet) that gravity obeyed an inverse-square law: Doubling the distance between bodies meant that the force between them was quartered. Hooke could not prove it, so Halley asked Newton, who professed to having known this all along—much to Hooke's irritation.

THE NEWTON

The unit of force is named the newton (N) in honor of the great physicist. 1 Newton of force accelerates 1 kg of mass at 1 m/s^2. The force of gravity acting on a falling apple (let's say it's about 100g) is approximately 1 Newton.

of gravity (F) equals on mass (M) multiplied by another (m) divided by the square of the distance between them (r). The gravitational constant (G) is a fixed conversion factor needed to make the calculation.

Laws of motion

The *Principia* also contained Newton's other contribution to the field of mechanics, his three laws of motion. These explain how a force—be it gravity, magnetism, or a simple push—interact with a mass to create motion. The first law reiterated the concept of inertia (developed by Galileo), by stating that a mass maintains its state of motion (including a lack of it) until a force is applied to change it.

The second law gives the magnitude of force as mass multiplied by acceleration, creating the simple but awesomely powerful formula: $F=ma$. This relationship tells you that a uniform force will accelerate a lighter object more than a heavier one, and to accelerate double the mass to the same speed requires double the force.

The third law states that when one mass exerts a force (an action) on a second mass, that second mass exerts the same force on the first (a reaction), only in the opposite direction. This explains everything from why rockets move to why we can't walk through solid walls—as you push through the wall, it will push back (hard).

Probl I

Investiganda est curva linea ADB in qua grave a dato quovis puncto A ad datum quodvis punctum B vi gravitatis suæ citissime descendet

Solutio.

A dato puncto A ducatur recta infinita APCZ horizonti parallela et super eadem recta describatur tum Cyclois quæcunque AQP rectæ AB (ductæ et si opus est productæ) occurrens in puncto Q, tum Cyclois alia ADC cujus basis et altitudo sit ad prioris basem et altitudinem respective ut AB ad AQ. Et hæc Cyclois novissima transibit per punctum B et erit Curva illa linea in qua grave a puncto A ad punctum B vi gravitatis suæ citissime perveniet. Q.E.I.

In Newton's day Latin was the language of European science, allowing people from different countries to communicate in a common tongue. In this Latin note, probably from the 1690s, Newton tackles the brachistochrone problem, finding the curve that allows an object to roll from one point to another in the minimum possible time.

THE LAWS OF MOTION

I

Every object in a state of uniform motion tends to remain in that state of motion unless an external force is applied to it.

II

The relationship between an object's mass (m)*, its acceleration* (a)*, and the applied force* (F) *is* $F = ma$.

III

For every action there is an equal and opposite reaction.

22 Theories of Light

ISAAC NEWTON DID MOST OF HIS RESEARCH IN SECRET, while lecturing on mathematics and physics at Cambridge University. Only later did he publish his findings. Even before Newton made his name with *Principia*, he had also already made great advances in the field of optics.

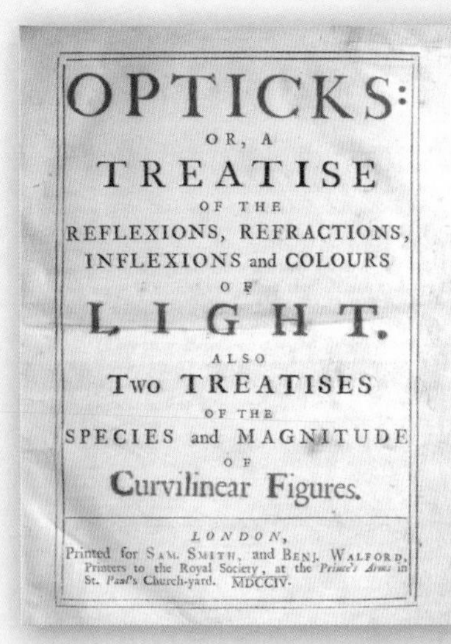

OPTICKS:
OR, A
TREATISE
OF THE
REFLEXIONS, REFRACTIONS,
INFLEXIONS and COLOURS
OF
LIGHT.
ALSO
Two TREATISES
OF THE
SPECIES and MAGNITUDE
OF
Curvilinear Figures.

LONDON,
Printed for SAM. SMITH, and BENJ. WALFORD,
Printers to the Royal Society, at the *Prince's Arms* in
St. *Paul's* Church-yard. MDCCIV.

The title page of the 1704 edition of Opticks. *Unlike the* Principia, *it was published in English, and was a more accessible read, being largely a description of Newton's optical experiments from the preceding decades.*

We have Newton to thank for the word *spectrum*, which he coined to describe the rainbow of colors that appears when white sunlight is split by a prism. In the early 1670s, Newton did this experiment himself but only published the results in the 1704 book, *Opticks*. Newton had also developed the reflecting telescope by this time, which used a mirror instead of a lens to collect starlight. The lenses of the day produced unwanted refractions, or chromatic aberrations, that split the light and created blurred images. Newton's telescope did not suffer from this problem.

In *Opticks,* Newton suggested that light was made of a stream of minute bodies, or corpuscles. This was at odds with the theory already proposed by Christiaan Huygens, who said light was made of waves. Newton rejected this because shadows have a sharp edge—and light waves would ripple around the edges. In the end, of course, they were both kind of right.

NAMING THE SPECTRUM

Newton came up with red, orange, yellow, green, blue, indigo, and violet for the colors of the rainbow. It would have been easier to use six colors, but Newton thought seven was a lucky number, so he invented indigo as the seventh shade.

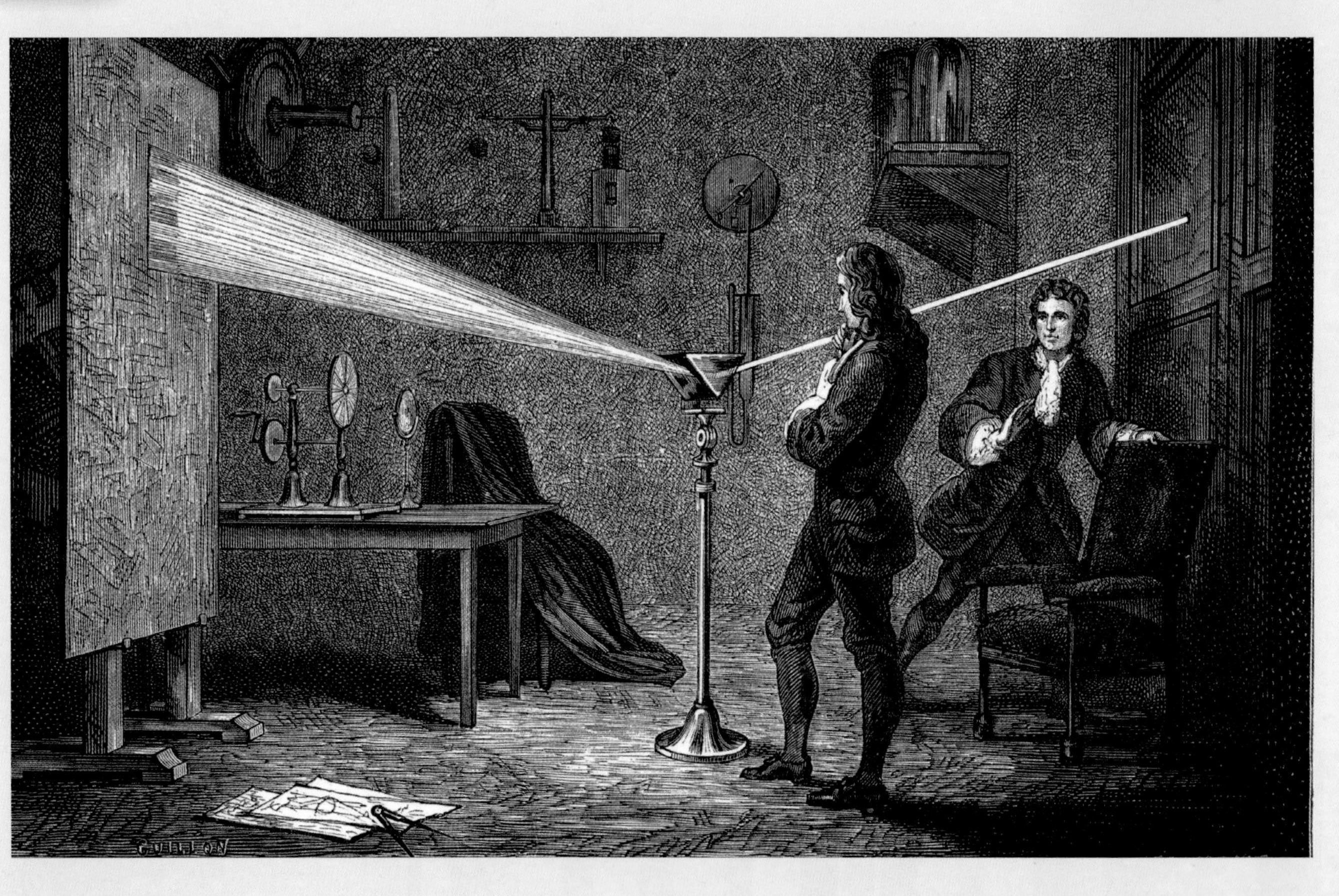

Newton shown in his darkened Cambridge drawing room with just a single beam of sunlight allowed in to refract through a glass prism.

23 The Flying Boy: Conducting Electricty

THE STORY OF ELECTRICITY DATES BACK TO THE VERY BEGINNING OF THE STORY OF PHYSICS, to Thales himself. However, until the 1720s, electricity was a static phenomenon. Then Stephen Gray made it move.

Otto von Guericke, the inventor of the vacuum pump did not just invent one groundbreaking device. He also invented an electrostatic generator that built up a charge by spinning a ball of sulfur. (It worked using friction, in the same way that rubbing a balloon will charge it.) New designs improved on this: Francis Hauksbee's even used a von Guericke vacuum pump to create an empty glass bulb in place of the sulfur ball to collect the charge.

Stephen Gray found that he could conduct electricity along wires. However, lacking enough metal, he used hemp "packing-thread" instead. Although not a fantastic conductor, he managed to send a current along a 240-meter (787-foot) length of it—after ensuring none touched the ground, or was "earthed."

Show business

Static electricity was certainly a crowd pleaser. "Electricians," performers who did tricks with sparks and static, were all the rage at society dinners. However, few scientists took an interest in this new supply of an age-old phenomenon. One of the few was Stephen Gray, a former cloth merchant, whose interest had been ignited by seeing sparks that occasionally formed on silk looms (again, the product of friction).

Thales had described how rubbing certain objects made them attract dust and feathers, and Gray did the same with long glass tubes. He found that he could extend the attractive force—or "electric virtue"—through a thread to an ivory ball, which collected fluff.

Taking his lead from other electricians of his time, in 1730 Gray performed a theatrical experiment, known as the "flying boy," to illustrate this further. He suspended a boy on silk ropes and conducted the charge from a Hauksbee generator to electrify him. The boy's hands attracted gold leaf, just like any charged object. The child was "in flight" off the ground to ensure that charge did not flow out of him to earth. The experiment showed that while electric virtue moved through metal, ivory, or the human body, materials like the silk ropes did not let it pass. The former substances are conductors, the latter insulators—a distinction that would prove crucial in understanding electricity and harnessing its power.

THE FIRST COPLEY MEDAL

Sir Hans Sloane became the president of the Royal Society, Britain's premier science club, after Isaac Newton's death. Sloane was Gray's friend (as well as the inventor of milk chocolate), and decided to honor him for his discoveries, after they were largely ignored by Newton. In 1731, Gray was awarded the Copley Medal, the highest accolade of the society, becoming the first of many eminent names to receive one.

24 Temperature Scale

IT IS FREQUENTLY THE CASE THAT SCIENCE IS FURTHERED NOT BY INTELLECTUAL ABILITY ALONE, BUT ALSO BY CRAFTSMANSHIP. This was the case with the thermometer, which was not invented, but was perfected by a German glassblower with a very familiar name.

The thermometer makes use of the principle that liquids expand as they get warmer and contract when chilled, something that Hero of Alexandria had already noticed in the 1st century CE. Early thermometers made use of water but were very inaccurate. In 1714, Daniel Gabriel Fahrenheit was skilled enough to make a thermometer containing mercury, which remains liquid when water froze or boiled. What was needed now was a scale to measure changes in temperature. Temperature scales are arbitrary and just need an upper and a lower point, both of which are easily repeatable so new thermometers can be calibrated. By 1724, Fahrenheit had chosen human body temperature as the upper point, set at 96°, and a mixture of ice, water, and ammonium chloride as the lower point of 0°. That made the Fahrenheit scale highly practical because it could be employed everywhere from a frozen peak to a boiling saucepan.

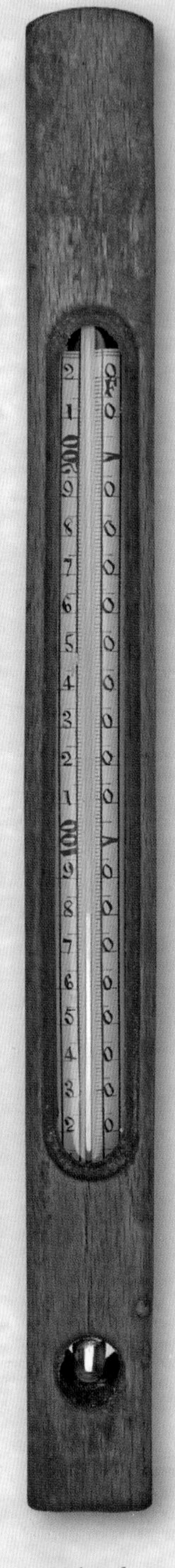

The quantity of mercury in the tube is small compared to the reservoir at the bottom. Fahrenheit's skill was in producing the precision glassware necessary for a scientific instrument.

25 The Leyden Jar

STEPHEN GRAY MAY HAVE SHOWN THAT ELECTRICAL "FLUIDS," AS THEY WERE DESCRIBED IN THOSE DAYS, COULD FLOW THROUGH CERTAIN SUBSTANCES, but they were decidedly hard to pin down. The search was on for a way of storing this marvelous phenomenon.

Although it would take another 150 years to figure this out, electricity is based on charged particles, chiefly electrons. A static charge is created by an object having a deficit or surplus of electrons. A spark is caused by those charged particles returning to equilibrium (while a current is a constant motion of them). In the 18th century, the perfectly reasonable explanation for static was that electrical "fluids" were pooling in an object (see box, left). Electrostatic generators, like Otto von Guericke's sulfur ball from the 1660s and Francis Hauksbee's glass sphere from the 1700s, were both "friction machines," that rubbed the fluid out of one solid into another.

Such electrostatic generators required considerable effort to produce only harmless

VITREOUS AND RESINOUS

In the 1730s, French researcher Charles du Fay found that there were two kinds of electrical "fluids": Resinous (found in things like amber) and vitreous (found in glass). Opposite electricities attracted each other, while like ones repelled. Today, these two states are relabeled as negative and positive charges.

sparks, and were of little use other than as novelties. One was the "electric kiss" party trick popular in the 1740s, where a couple literally created a spark between them—once one had been electrically charged. To do more with electricity, one would need to capture a larger quantity. And where better to collect a fluid than in a jar?

Dangerous experiments

In 1745, a German scientist called Ewald Georg von Kleist lined the inside of a glass jar with silver foil, and filled it with some water. The idea was to charge the water by connecting it to an electrostatic generator. His idea was flawed but something was going on, because when he touched the foil with his hand, the electric shock was very powerful—and certainly very dangerous. However, Kleist survived. His jar had obviously stored electricity. But how did it do it?

Von Kleist gets a helping hand to charge his prototype Leyden jar. He is in for a bit of a shock. Later developments showed that the jar still worked perfectly when the water was left out.

A similar device was built by Dutch inventor Pieter van Musschenbroek—he had had the same teacher as von Kleist. Van Musschenbroek showed the jar to academics at the University of Leiden, and the apparatus was named the Leyden jar. Improvements were made. Foil was added to the outside as well as the inside, although a gap was left at the rim of the jar. The second foil took the place of the hand of the electrician, which had been a working part of the earlier device.

According to the primary account, the American statesman Benjamin Franklin confirmed that lightning strikes contained "electrical fluids" by collecting charge with a kite tied to a key—until sparks jumped from the key to a ring on his finger! It is suggested that he also charged a "phial" with the key, often interpreted as being a Leyden jar. Whether he actually did such a foolhardy experiment remains a mystery.

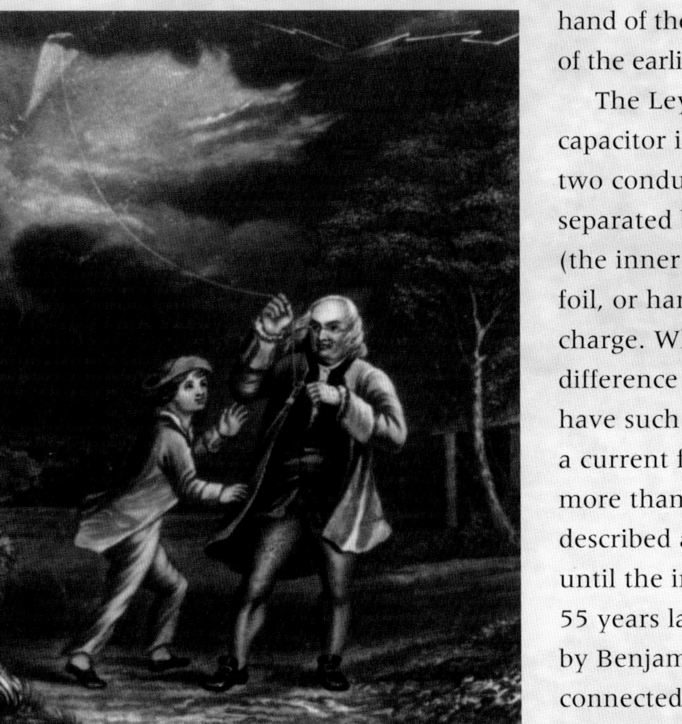

The Leyden jar was the first capacitor. A capacitor is a device for storing charge made of two conducting plates. The plates (the foils) are separated by an insulator (the glass). As one plate (the inner foil) is charged, the second plate (outer foil, or hand) builds up an equal and opposite charge. When the two plates are connected, the difference in charge is rectified. Modern capacitors have such a large surface area that they will release a current for a short while. Leyden jars did little more than produce large sparks—too brief to be described as currents. That would have to wait until the invention of chemical batteries around 55 years later. However, the word *battery* was coined by Benjamin Franklin, because banks of Leyden jars, connected up to work together, reminded him of a battery of cannons.

26 Hidden Heat

Joseph Black was a Scottish doctor with a wide range of interests. He had already added chemistry to that list when he stumbled upon carbon dioxide while researching stomach remedies, but his biggest contribution is to physics with the discovery of "latent heat."

In the 1750s, Black used accurate mercury thermometers to track how heating a substance changed its temperature. He discovered that different materials warmed up at different rates, a phenomenon he called "specific heat." A decade earlier Swede Anders Celsius had devised a temperature scale based on the melting (0°) and boiling (100°) points of water, a rival to Fahrenheit's to this day. Black found that heating melting ice does not increase its temperature but results in more water. The same was true of heating boiling water—the result was steam of the same temperature. Black concluded that heat was being used to transform the state of a substance, rather than merely making it warmer. Only when the melting or boiling were complete did the addition of heat again lead to a rising temperature. Black called this latent heat: The latent heat of fusion is needed to melt a solid, while the latent heat of vaporization is concerned with evaporation.

Joseph Black (seated) discusses his ideas about heat with colleagues, one of whom is James Watt (center), the engineer who went on to harness heat to drive the first viable steam engines.

27 Fire and Matter

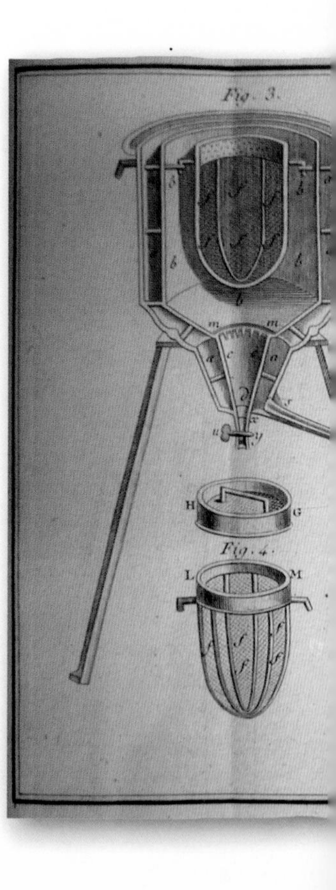

Mercury thermometers made it possible to measure how hot something was, but a question remained: What is heat? Even in the brightest part of the Age of Enlightenment, the influence of ancient physics was still being felt as science searched for an answer.

Joseph Black's research was based on the theory that heat was a substance, a thick liquid that he called *caloricum*, which flowed, somewhat stubbornly, from one substance to another. This thinking was the result of many centuries of belief that fire was a substance. In the Middle Ages, it was suggested that sulfur was some kind of fire in solid form. By the 18th century, the dominant theory was that substances burned because they contained phlogiston—the heat and light of a flame was the result of phlogiston leaving the substance. However, the evidence suggested something different. According to the phlogiston theory, hot metal glows as phlogiston is released. However, heating certain metals actually makes them heavier, suggesting something was being added (a layer of oxide in this case).

Combustion revealed

Phlogiston was consigned to history by Antoine Lavoisier, a gentlemen chemist from France. In the 1770s, he showed that burning, or combustion, was substances reacting with oxygen—the now well-known gas then only recently discovered. It was Lavoisier who named the gas, as he had done for hydrogen. Burning both these gases together created water—initially as steam but it cooled into a liquid. Lavoisier's experiments also confirmed what Russian Mikhail Lomonosov (see box) had predicted 20 years before: The weight of hydrogen and oxygen gas before a procedure was the same as the liquid water produced. This proved that matter was not created or destroyed, just rearranged. There was no place for phlogiston at all.

HEAT IS MOTION

Mikhail Lomonosov, an 18th-century Russian polymath, contributed to many sciences but remains an unsung figure outside Russia. Others had suggested that heating gas resulted in it moving faster, and Lomonosov turned that idea around. He theorized that all heat, wherever it was found, was the result of motion within the fabric of a substance. Our modern understanding is based on the same idea.

Russian empress Catherine the Great pops in to visit Lomonosov, as he shows off some of his scientific apparatus.

Measuring heat

Lavoisier supposed that heat and light were two kinds of the same thing. Since Newton had said that light was made of corpuscles, or tiny weightless particles, Lavoisier decided that heat was as well. He added a Gallic twist to Black's Latin word, naming the material *calorique*.

In the 1780s, Lavoisier partnered with Pierre-Simon Laplace, another French scientist with an intellect at least as colossal as his own. They set themselves the task of building a machine for measuring quantities of calorique—in other words a calorimeter. The device they came up with was a combustion chamber that was isolated from any influence from the outside world by insulating layers. The central chamber itself was submerged in a precise volume of tightly packed ice crystals. The calorique flowing out from the chamber would melt a portion of the ice. The exact proportion of the ice and meltwater that resulted was a measure of calorique released. As the word suggests, the calories in food are measured using modern forms of this device.

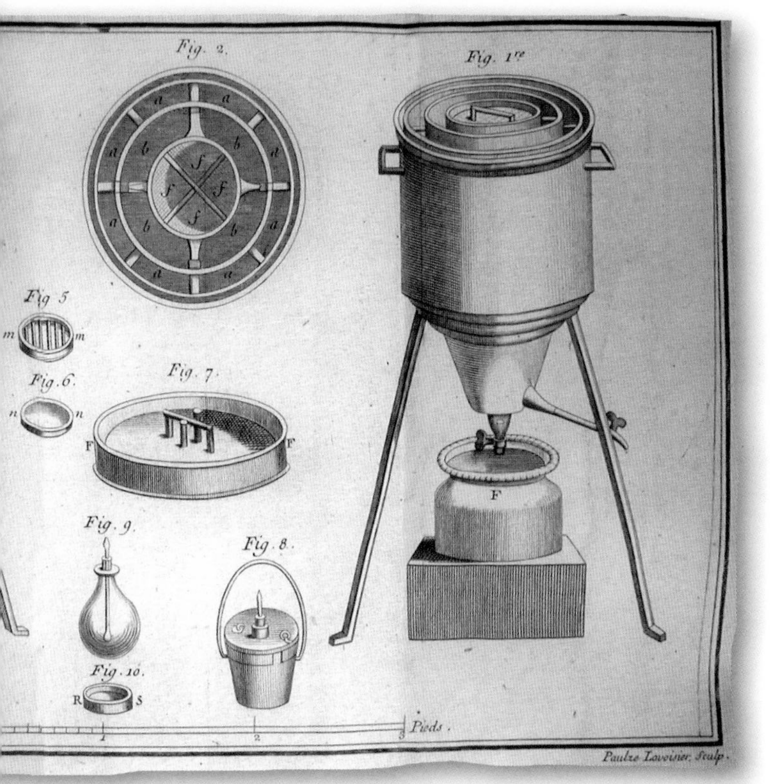

A diagrammatic plan of the Lavoisier–Laplace calorimeter, complete with removable combustion chamber, lid, and drainage system to collect the water melted in the experiment. It is reported that a guinea pig was put in the device to show that the heat from a body was the very same as that released by burning.

28 Measuring Charge

IN THE 1780S, A FRENCH PHYSICIST DEVISED A WAY OF MEASURING the forces exerted by electrified objects. It showed up a familiar relationship.

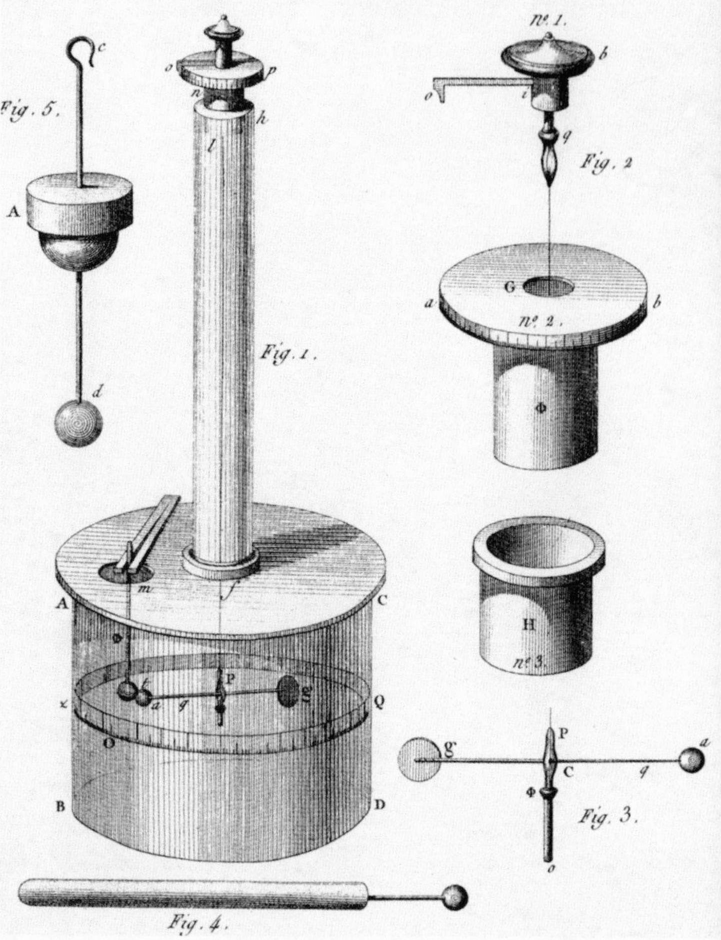

A diagram of Coulomb's torsion balance. The angle of the bar's swing was used to measure the size of the force exerted.

Pioneers in the field of electricity had developed various kinds of electroscopes, detectors that picked up the forces coming from electrified objects. However, because the forces involved were so weak they were ill designed to quantify them. In 1784, Charles-Augustin de Coulomb devised the torsion balance, which was sensitive enough to measure forces of this magnitude. It had a metal bar that hung from a thread so it was free to swing under the lightest of forces. Once it was charged, the bar moved whenever another charged object was placed nearby. Coulomb found what is now known as Coulomb's law— the size of force is inversely proportional to the square of the distance between the charged objects. This is a relationship that electrical phenomena share with another natural force, namely gravity.

29 Weighing a Planet

NEWTON HAD SHOWN HOW THE GRAVITY BETWEEN OBJECTS IS RELATED TO THEIR MASSES AND DISTANCES. Within that relationship is "big G," a constant of proportionality, an unchanging number that defines the Universe, and which allowed Henry Cavendish to weigh the world.

Newton's work on gravity proved the unimaginable: As an apple falls to Earth, Earth moves up to meet it. The huge disparity in the pair's respective masses means that Earth barely moves at all, but Newton's laws of motion states that it *does* move.

To calculate how much Earth moved, one would need to know the mass of the planet. No easy feat. However, in 1789, a great experiment was devised by English scientist Henry Cavendish to weigh Earth. The law of gravitation ($F=G(m_1m_2)/r^2$) could do the job. All that was needed was a figure for the constant of proportionality. This is the number that underwrites the link between the force of gravity with mass and distance; today, we know it as "big G." ("Little g" refers to the acceleration due to gravity.) G is a universal constant, the same number works for any mass, large or small. The Cavendish experiment used one of each.

Torsion balance

Henry Cavendish built a torsion balance—a detector that responds to force with a twisting action. Cavendish's balance was built to detect forces even smaller than Coulomb's electrostatic device could, and so it was built on a grand scale—enough to fill a shed on Cavendish's estate. Two identical lead balls, weighing 158 kg (348 lbs), were suspended from a beam, so they could be swung into position. They were each placed next to smaller lead balls, weighing 730 g (25 oz), which were suspended on a separate system that was free to swing independently.

The gravitational pull between the balls made the apparatus swing—most motion was made by the smaller masses. The swing of these smaller balls only stopped when the force of attraction was balanced by an equal but opposite twisting force, or torque, of the wire from which the balls were suspended. Cavendish knew the torque for any given angle the device rotated, and that figure allowed him to calculate G. Converted to modern units his figure was 1 percent out from the current value of 6.67259×10^{-11} Nm^2/kg^2. Cavendish then used the acceleration due to Earth's gravity (g=9.8 m/s^2) to calculate Earth's density, which he found was a little over five times that of water. This technique let him bypass calculating a weight for Earth directly (the density was a good starting point), but in doing so he had revealed a more significant number: G.

Henry Cavendish shielded his balance from outside influence. As this replica shows, he used a pulley to position the large balls and watched the motion of the small ones with telescopes through the walls.

30 Frogs Legs and Piles

THE BREAKTHROUGHS THAT UNLEASHED THE POTENTIAL OF ELECTRICAL CURRENTS COULD NOT HAVE COME FROM MORE DIFFERENT SOURCES. One was an accidental discovery by a doctor studying frog anatomy, the other was a deliberate attempt to make electricity using a chemical reaction.

Luigi Galvani was an unlikely physicist. He had followed his father into the medical profession. As part of his training in surgery he developed an interest in anatomy—choosing to study the internal structure of dead animals before being let loose on a living patient. Eventually, he became a full-time anatomist at the University of Bologna in Italy. After nine years of academic research, Galvani made the discovery for which he is remembered by chance. He had hung up pairs of frogs' legs on a wire fence to dry out. The fence was iron while the hook was copper, and the fresh frogs' legs started to twitch—even spark according to some accounts!

Galvani found he could recreate the twitching with electrical charge from a Leyden jar, showing that living (at least recently dead) muscle was stimulated by what he termed "animal electricity." However, Galvani's investigations into the original phenomenon on his fence had wider implications for physical science. He recreated the situation in the laboratory, using a metallic arc made from the same two metals—copper and iron—to connect the tip of the frog's toe to a dissected spinal cord,

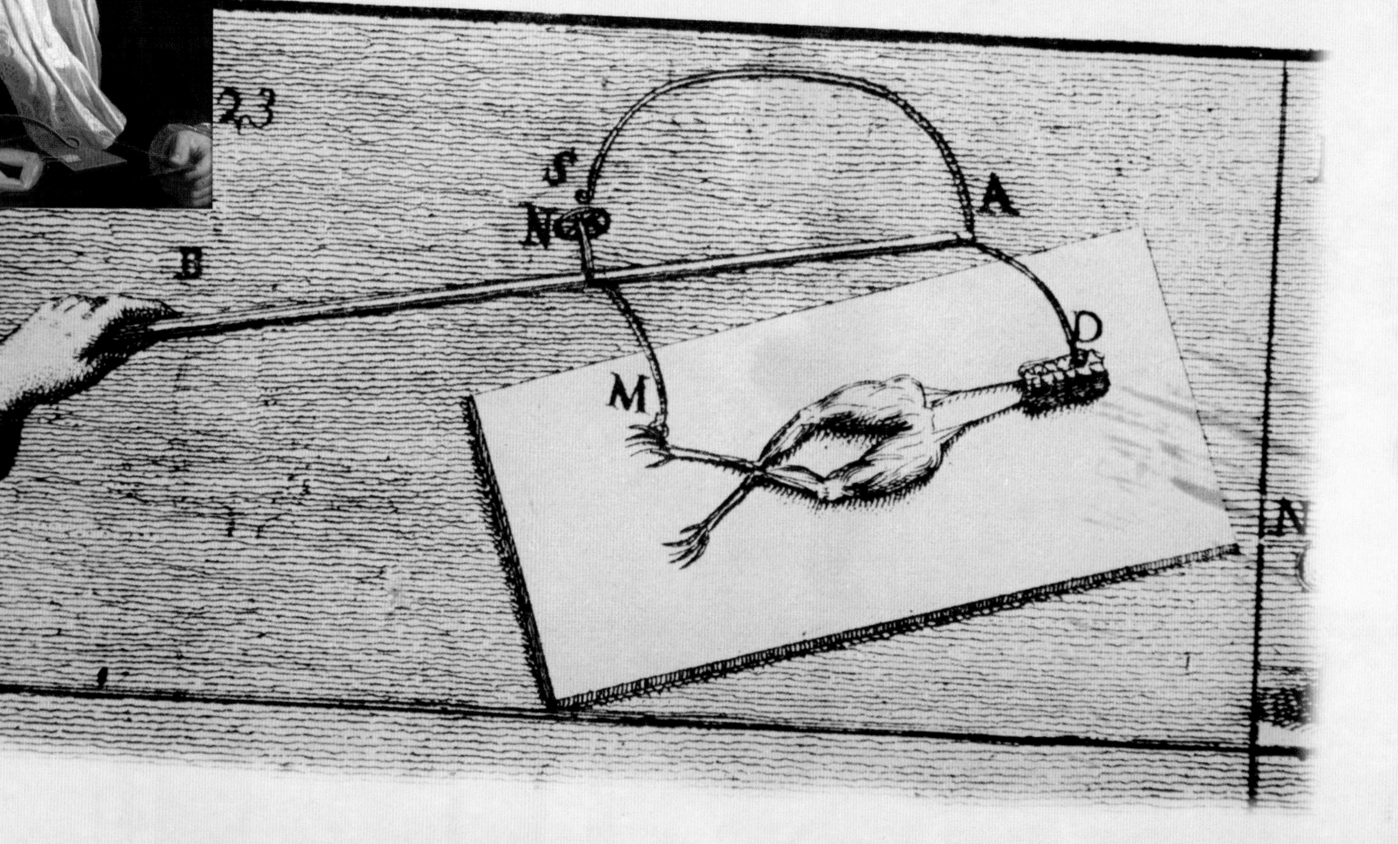

Although he did not know it at the time, Luigi Galvani's metal-meets-frog apparatus was the first example of a battery.

Alessandro Volta shown with his "voltaic pile" at the turn of the 19th century. As his name suggests, the electrical unit volt is named for this individual. A volt is a measure of the force that pushes electric currents through circuits.

FRANKENSTEIN

Galvani's nephew Giovanni Aldini turned animal electricity into a show. He toured Europe zapping the corpses of recently executed convicts so they quivered and twitched—to much acclaim. Mary Shelley, the author of **Frankenstein**, a story of a monster reanimated by electricity, is said to have been inspired by Aldini's macabre performances.

Dr. Frankenstein prepares to electrify his monster in a 1994 movie.

which contained the nerves that controlled the leg muscles. In so doing he had built an electrical circuit that allowed "animal electricity" to flow through the leg muscle, making it twitch. But where did the flow of electrical charge come? Galvani believed that he had found some vital force specific to living things, but 30 years later another scientist showed how to produce the same effects without animals.

Electrical pile

Galvani's metallic arc only worked with fresh meat, still oozing with body fluids. Italian Alessandro Volta replaced the flesh with wood pulp soaked in salty water. He had realized that the significant components were the two metals, which were reacting with each other and somehow causing electricity to flow from one to the other. Volta maximized the effect by repeating the bimetallic units many times, literally piling them on top of each other. His first "voltaic pile" alternated between silver coins and disks of zinc, all separated by soggy wood pulp. Wiring the top of the pile to the bottom allowed the electricity to flow. So Galvani's animal electricity was the same thing as Volta's "heat electricity," as he called it. To figure out why, physics would have to discover more about what nature was made of.

BATTERY

A modern battery works on the same principles as a voltaic pile. Two substances (the anode and cathode) are made to react with each other. As they do so, electrons are passed from the anode to the cathode. The battery is designed to keep these two reactants apart, so the electrons have to flow between them through a liquid called the "electrolyte"—thus creating the electric current.

Positive terminal
Anode (Zinc inner case)
Cathode (Graphite rod)
Electrolyte paste
Negative terminal

31 Atomic Theory

IN 1803, AN ENGLISH SCIENTIST DID SOMETHING UNUSUAL. INSTEAD OF DISPROVING THE LONG-HELD THEORIES ABOUT THE UNIVERSE from ancient Greek philosophers, he showed that one of them was actually right. Democritus's atoms turned out to be a fundamental component of matter.

By the turn of the 19th century, "pneumatic" scientists had revealed that air was in fact a mixture of gases—such as carbon dioxide, nitrogen, and oxygen. They had also found that chemical processes produce a range of other gases, such as hydrogen—and burning hydrogen and oxygen together produced water. So the classical view of natural substances had come crashing down. There were dozens of elemental substances, with new ones being found all the time. Earth, water, and air were not elements at all—and the evidence suggested that fire was just hot, glowing gases.

John Dalton spent most of his working life in Manchester, England, a city where many of the later breakthroughs in atomic theory would also arise.

Mixture of gases

It was against this backdrop that Englishman John Dalton began to think about the characteristics of gases. In 1738, Daniel Bernoulli had produced some mind-bending math that expressed the pressure exerted by a gas as a series of theoretical points of mass in constant motion and giving a little push against the inside of the container. However, Dalton arrived at the problem from a different direction—weather forecasting.

In his early twenties, Dalton began recording meteorological conditions—a practice he kept up until his death in 1844. The daily data included the varying air pressure, which had been related to changing weather since the days of Blaise Pascal.

As Dalton puzzled out the relationships between the air and weather, he came to a bigger realization. He was aware that the two later gas laws (Charles's and Gay-Lussac's) had recently been added to Boyle's to describe the behavior of gases. Air had also been established as a mixture of gases, a mixture that was in constant flux—evidenced by the changeable English weather. Dalton's first contribution was to suggest that the total pressure of the air could be divided up into "partial pressures" exerted by the different gases in the air. This concept is known today as Dalton's law.

DIFFUSION

Most of the gases discovered in the early years of pneumatic chemistry were colorless, but not all. Nitrogen dioxide, for example, is bright orange—and has a strong smell. Such a gas showed scientists that gases diffuse—unerringly spread out until they fill their container evenly, no matter how large the volume.

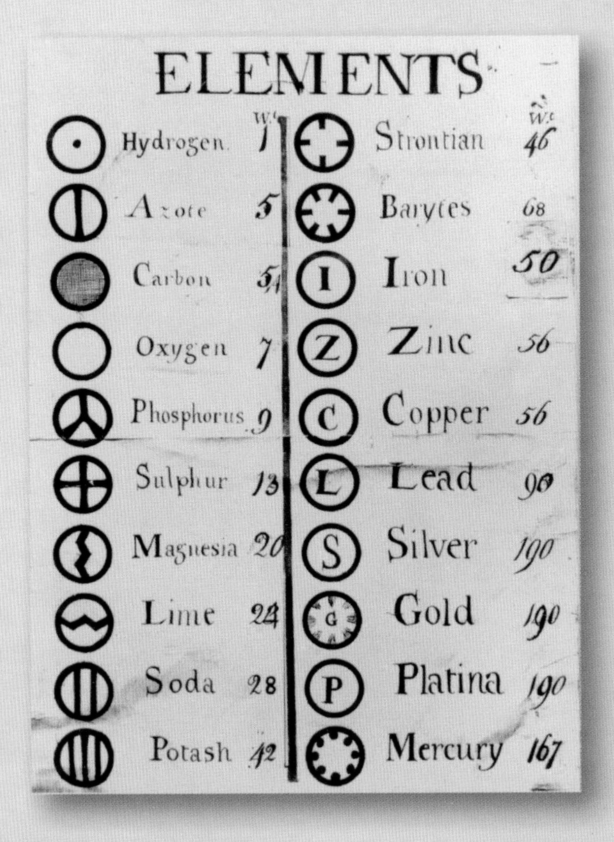

Dalton's table of elements is arranged by atomic weight, with the weights of heavier substances shown as multiples of hydrogen's. The weights are largely inaccurate and several items are not elements at all. But despite these errors, Dalton's table shows a major step forward in science.

Spreading out

Dalton's law also points out that the gases in a mixture diffuse independently of each other, so they continue to exert their partial pressure in all parts of a container. In other words, if two pure gases are mixed in equal volumes, a sample of the mixture will always be half one gas and half of the other. This showed Dalton that whatever it was that gave gas its physical characteristics—and exerted pressure—every gas was made up a unique type of material.

As well as physical properties, the units of gas also exhibited chemical ones. Dalton performed many experiments to analyze how elements such as hydrogen and oxygen combined to form compounds like water. These confirmed that elements combined in fixed ratios, always whole numbers. For example, carbon and oxygen combined in a ratio of 1:1 to make carbon monoxide. Burning this product formed carbon dioxide, where the carbon–oxygen ratio was 1:2. This was Dalton's "law of definite proportions."

The same experiments showed Dalton that some gases weighed more than others. Hydrogen and oxygen looked identical, but a flask of the latter was considerably heavier than the former. Here was the final piece of evidence. In 1803, Dalton proposed that gases were made from invisibly small particles, and reaching back to antiquity he called them "atoms," as they had been named by the ancient Greeks. Since gases could change into liquids and solids, and back again, the implication was that everything was made from these atoms. Dalton's theory stated that atoms of one element were identical to each other but different from those of other elements. For physicists, the question now was: What are atoms made from?

CHEMICAL MOLECULES

Dalton described a chemical reaction as the process by which atoms are combined, pulled apart, or rearranged. He reimagined the fixed proportions he had discovered as atoms of different elements bonding together to form clusters with particular geometries. A word for them had been coined a few years before—"molecules." A molecule is to a compound what an atom is to an element, the smallest possible part. If either is broken up, they cease to represent the same substance.

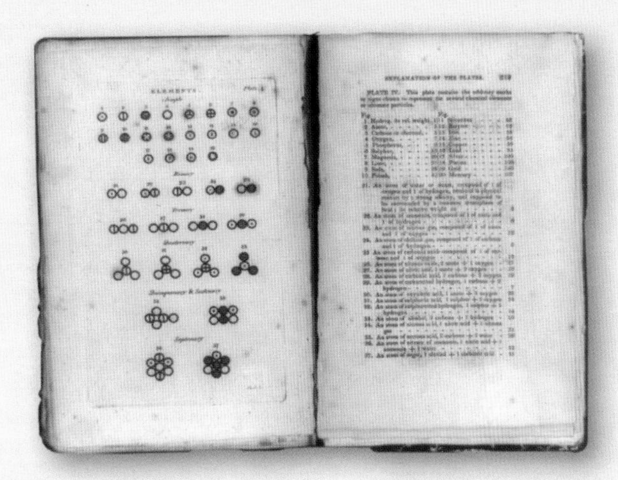

Dalton represented molecules as diagrams and built model atoms out of wood to visualize the ways they could bond.

32 Light is a Wave

WHILE ELECTRICITY AND HEAT WERE NOT DIRECTLY OBSERVABLE—EVERYTHING WAS SURMISED FROM THEIR EFFECTS—light could at least be seen. By the 19th century, two theories about it were firmly entrenched. And to question either of them meant to have your loyalty questioned in turn.

For the whole of the 18th century, investigations of the nature of light had been clouded by nationalistic tensions. In 1678, Dutchman Christiaan Huygens had proposed that light was a wave. Thirty years later, the English scientific behemoth Isaac Newton dismissed this, putting forward a "corpuscular" theory: Light was a stream of particles. Newton liked this version because it allowed him to treat light like other materials on the move—with minute bodies ricocheting around the Universe according to his laws of motion.

By the time Newton had released this theory, he was the loudest scientific voice in the world. He did not suffer contradictions easily on this or a number of other issues where European scientists offered different views. The sheer force of this side of his personality continued after his death, and meant that English-speaking scientists were all expected to follow Newton's teaching.

This ripple pattern is created by light beams interfering with each other, conclusive proof that light behaves like a wave.

Spreading waves

In continental Europe, the opposite was the case. A few opted for René Descartes' pressure theory, but Huygens's approach was the leader. It pictured light as a periodic oscillation that propagated in all directions from a source of light and at a particular speed. He developed geometry that showed how light waves behaved when they met an obstacle. He described the wave front as a series of points that could each become a source of light, and send out new "wavelets" in all directions. This theory could be used to explain many of the observed behaviors of light, some of which Newton's theory could not—most crucially the way two light beams interacted with each other.

Still in his 20s, English doctor Thomas Young was perhaps youthful and brave enough to go against the Newtonian view. (Benjamin Franklin was also in the wave camp.) Any chorus of disapproval over Young's research soon faded because by 1804 he had offered two demonstrations that proved that Huygens was right.

The wave theory of light predicted that light could be made to behave like the ripples in water.

Water meets light

The first experiment was to observe the behavior of waves of water in a ripple tank. Young made two wave fronts collide—or interfere—and showed how they could combine into larger waves or cancel each other out. He also sent a straight water wave through a small gap. This showed, as Huygens described for light, that the small part of the wave that could get through propagated in all directions once it arrived on the other side of the gap.

Then Young performed what is known as the Young experiment, sending beams of light through gaps (see box). The patterns of light that emerged on the other side showed that light behaved just like the ripples of water.

YOUNG'S INTERFERENCE EXPERIMENT

Also known as the double-slit experiment, this demonstration is a staple of physics classes the world over. The first slit is used to create a point source of light with a single wave front. This then hits two more slits and forms two points of light, which propagate in all directions (unless blocked by the screen). The propagating waves collide with each other and interfere. Waves have peaks and troughs, and when two waves collide they add together. A peak added to a peak creates an even higher wave (or brighter light). The same is true when troughs meet. However, if a peak hits a trough, they cancel each other out, and the wave disappears—producing darkness. In the diagram below the red lines represent wave peaks, the yellow ones are troughs. As you can see the interfering light creates a pattern of light and dark lines—proof that light is a wave.

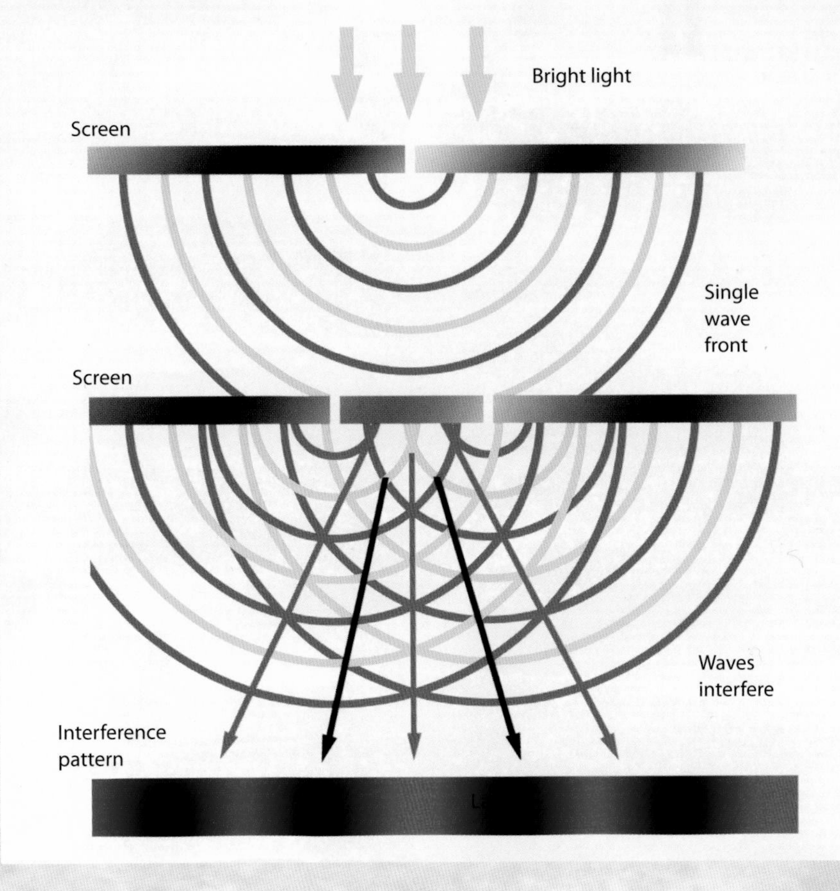

33 Plastic and Elastic

HOOKE'S LAW FROM THE 1660S RELATED FORCE TO STRETCHING, AND EACH MATERIAL HAD A PARTICULAR STIFFNESS. "Young's modulus," named for Thomas Young, is a way to quantify that stiffness.

Rubber bands have a very low Young's modulus, so small forces lead to long elastic extensions. A band will spring back to its original length—or break!

The idea for what is now called the Young's modulus occurred to the Swiss mathematician Leonhard Euler in the 1720s. It is a number particular to every substance and has to be measured experimentally. Young published many results in 1807 and the numbers have carried his name ever since. A Young's modulus is calculated as the tensile stress divided by the strain. Stress is the force per unit area pulling down on the object. Strain is the proportion that the object stretches. The Young's modulus holds until a substance stretches past its elastic limit. After that, increased stress will create permanent, or plastic, deformation—and eventually break it. Steel has a high Young's modulus, which means it is stiff and does not stretch much. Rubber has a low Young's modulus.

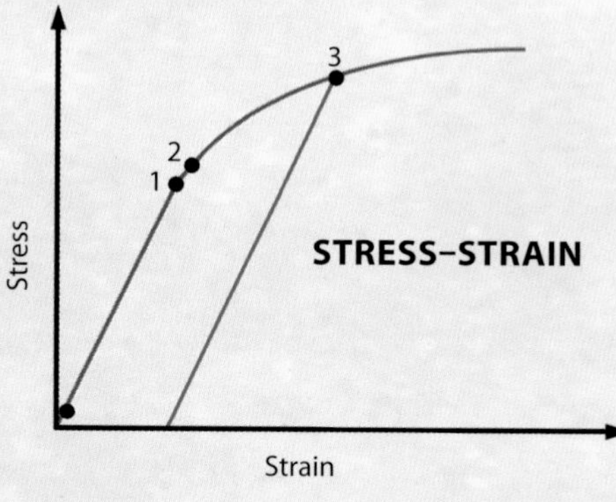

This stress–strain graph shows how a typical substance stretches. The straight section is where Hooke's law applies. Point 1 is the limit of proportionality, after that strain is no longer proportional to stress. Point 2 is the elastic limit. If the strain is removed at Point 3, the object springs back to a shorter length, but has become permanently stretched.

34 Electricity Meets Magnets

THE ABILITY TO HARNESS SEEMINGLY ENDLESS AMOUNTS OF ELECTRICAL ENERGY IS ONE OF THE FOUNDATIONS OF THE MODERN WORLD. That technology is built on an area of physics called electromagnetism—a field that came into existence by accident.

Until 1820, electricity and magnetism were studied separately. It was thought that only iron could act as a magnet. Artificial magnets, such as compass needles, were made either by painstakingly rubbing a lodestone along them repeatedly or by aligning them north to south and gently heating and tapping. Electricity was something completely different, involving the transfer of a charged fluid as sparks or in continuous currents from the recently developed voltaic pile and other chemical-based batteries. However, links between the two phenomena were beginning to be considered: For example, the iron knives in houses struck by lightning were frequently left magnetized.

Ørsted shows off his electromagnetic discovery using a Daniell cell, a battery powered by reactions between two electrodes separated by a salty liquid.

Chance discovery

Hans Christian Ørsted, a professor at the University of Copenhagen in Denmark, was one of the last natural philosophers in the true sense of the word, making contributions to both science and philosophy. One of his fields of expertise was the work of Immanuel Kant, a German philosopher who saw scientific phenomena as different facets of the same underlying natural order. Ironically, Ørsted's greatest scientific achievement was to show just that.

In April 1820, Ørsted was giving a lecture on the heating effect that electric currents had on metal wires, using a voltaic pile to demonstrate. A compass needle was on the desk for use in a later lecture. When the current was on, Ørsted noticed that the needle swung to point at his hot wire. When he switched the current off, the compass needle returned to the north. Ørsted immediately saw what it meant: The electrical current was making the wire into a temporary magnet. (Today, we call such devices electromagnets. They are very useful magnets that can be turned on and off.)

Ørsted thought that heat was the crucial link, and that magnetic force was radiating out of the wire like heat or light. He worked with thinner wires (which get hotter) but the magnetic effects remained feeble. Nevertheless, the link had been made, and the field of electromagnetism was born.

Ampère repeats Ørsted's experiment in his lab in Paris.

AMPÈRE

Ørsted's find confirmed the research of Frenchman André-Marie Ampère. Within months Ampère had presented a clearer picture of electromagnetic forces. He showed that two electrified wires behaved in the same way as one wire and a magnet. He also found the polarity of the magnetic force from a wire depended on the direction of the current. Currents running in opposite directions caused a repulsive magnetic force, while currents running in the same direction attracted each other. The unit of current is named the amp, short for ampere in his honor.

35 The Thermoelectric Effect

A FURTHER CHANCE DISCOVERY INVOLVING HEAT AND ELECTROMAGNETISM led to one of the central laws of the field.

Thomas Seebeck was a researcher at Berlin University who was looking into the magnetization of various metals using heat. In 1821, he found that when he heated one end of a loop of two different metals, it became magnetic. Ørsted advised that this was a consequence of an electric current running through the loop. The temperature difference was creating a potential difference—an imbalance in charge—that was forcing current to move through the conductors. (This is better known today as a voltage.) Seebeck had discovered the thermoelectric effect, where heat energy becomes electrical energy (and vice versa). Another German used this effect to study the relationship between voltage (V), current (I), and resistance (R, a measure of how easily a current moves through a conductor). His name was Georg Ohm, and the law he discovered, $V=IR$, bears that name (as does the unit of resistance, the ohm [Ω]).

A replica of Seebeck's metallic loop. The compass needle inside aligns itself to the metal strip when heat is applied at one end—the so-called Seebeck effect.

36 Heat Engines

AS PHYSICISTS WERE MAKING THESE SMALL BUT SIGNIFICANT STEPS FORWARD AT THE START OF THE 19TH CENTURY, engineers were taking huge strides. Immense steam engines were putting heat to work. Their mechanisms were highly refined, but the theory behind them needed work.

Steam technology was centered on Britain, where engines built by the likes of James Watt and Matthew Boulton were helping to create the world's first industrial nation. So when Sadi Carnot, a young French soldier, began to study engines, few took any notice. However, his work would not only prove crucial for making better engines, but also throw light on the nature of energy itself.

Carnot's father had been a somewhat tyrannical French revolutionary, and that hindered his son's military career. Instead, Carnot devoted himself to

Sadi Carnot presented his research into heat and energy in an 1824 book called Reflections on the Motive Power of Fire. *He did not live to see the impact of his work. In 1832 he was admitted to an asylum suffering from mania and delirium and died shortly after at the age of 36. However, beyond the grave he was dubbed "father of thermodynamics."*

research into steam engines. Steam engines function by boiling water into hot steam, which does "work" as it expands, pushing against mechanical parts, making them move. Carnot was interested in how heat could become motion and imagined four steps occurring inside a theoretical engine, now known as the "Carnot cycle." In stage one a hot gas expands, doing work without losing any heat. In stage two, the expansion continues, but the gas cools as its heat energy is converted to motion. For stage three, the gas is compressed by the machinery but does not warm up, and in stage four, further compression heats the gas (in accordance with Charles's law). Carnot's idealized "heat engine" shows how heat can do work and how work can cause heating: The work done by his theoretical engine was equivalent to the temperature change of the gas inside. Could an engine actually do this?

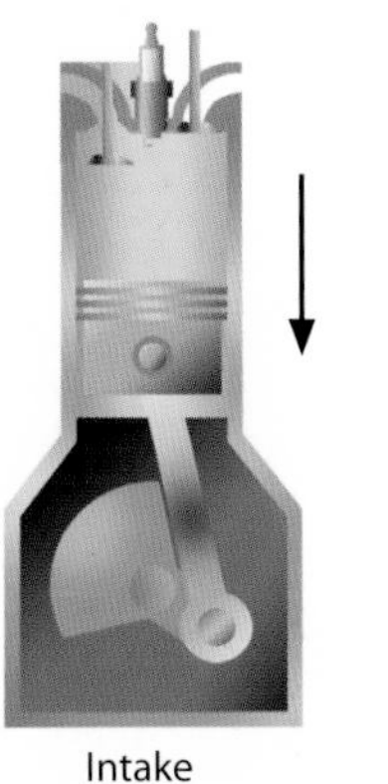

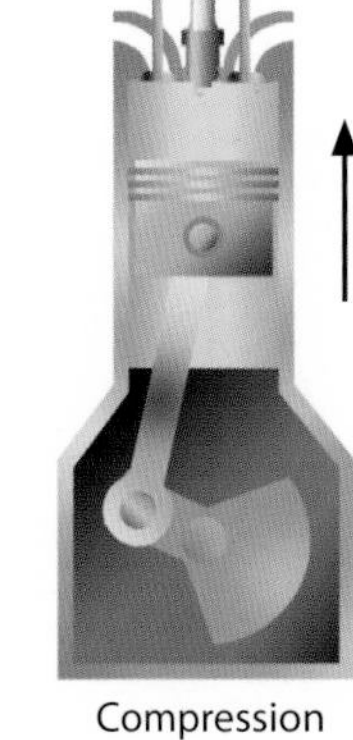

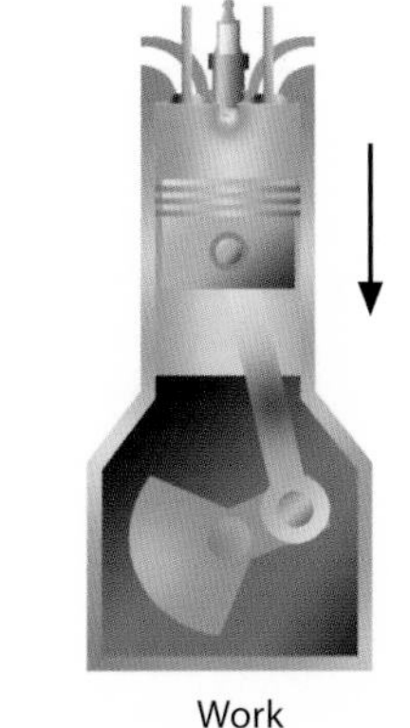

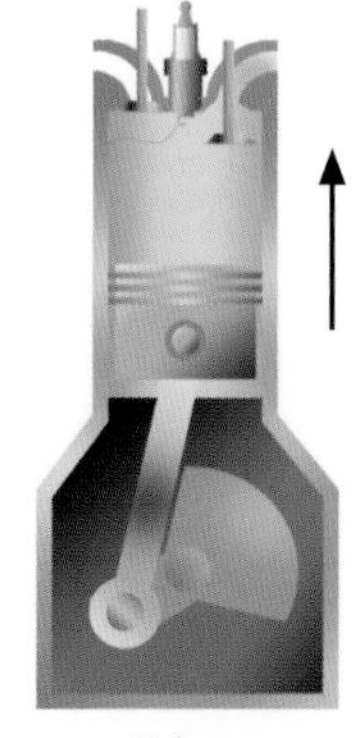

The four-stroke car engine uses a refined version of the Otto cycle, named for German Nikolaus August Otto. This builds on the Carnot cycle but uses the heat from combustion inside the moving parts—hence, it is an internal combustion engine.

37 Brownian Motion

As the only botanist in our story, Robert Brown's physics discovery came from a most unusual source, from peering down a microscope at pollen. What he saw became the first visible proof that atoms actually exist.

Botany was a cutting-edge science in the early 19th century. Joseph Banks, a botanist-cum-explorer, presided over the Royal Society of London, and like him, Robert Brown spent years abroad gathering specimens. In 1827, he was examining the pollen of ragged robin, a pink wildflower from the Pacific Northwest, under a microscope. He noted tiny particles (vesicles of starch and oil) were released into the water and began to joggle around in all directions. In 1785, Jan Ingenhousz had seen the same effect in coal dust suspended in alcohol and commented how lifelike the specks appeared. Brown said the opposite—the motion had a physical, not a biological, source. Ninety years later Albert Einstein explained that the so-called "Brownian motion" was caused by vibrating atoms and molecules that were in constant collision with the tiny objects.

Tracking the paths of particles in Brownian motion results in truly random squiggles. The motion of the large visible particles is being directed by a multitude of fast-moving particles, too small to see. Nevertheless, analysis of the motion gives proof that they are there.

38 Inducing Currents

Michael Faraday gives a lecture on his discoveries at London's Royal Institution in 1856. Faraday had begun a tradition of Christmas lectures for the general public in 1825. They are still given by top scientists each year and broadcast around the world.

AS THE LINK BETWEEN MAGNETS AND ELECTRICITY BECAME ESTABLISHED, people looked at ways of harnessing their forces to create motion. The man who succeeded first also found he could use motion to create electric currents.

The eminent English scientist Humphry Davy and his colleague William Wollaston were two of the leading experts in electricity. In 1807, Wollaston had constructed a huge battery of voltaic piles in the basement of the Royal Institution in London, an upstart research center to rival the Royal Society across town. Davy used the electrical energy from the pile to split materials into their constituents—discovering five new elements, such as sodium, potassium, and magnesium in the process.

In 1813, an apprentice bookbinder barely into his twenties came to hear the great Davy speak on the subject of electrical research. The young man was called Michael Faraday. He wrote up some notes on the lecture, which so impressed Davy that he invited Faraday to be his assistant.

Motor run-in

Faraday joined in Davy and Wollaston's quest to build an electric motor that turned the repulsive and attractive forces of electricity and magnetism into rotary motion. In 1821, armed with Ørsted's discovery of electromagnetism, Faraday produced the world's first electric motor, and he did it independently of his mentors, much to their personal fury.

Faraday was forced to limit further electromagnetic research and keep it

HENRY AND SELF-INDUCTANCE

The story of electromagnetic inductance does not begin and end with Michael Faraday. In the same year as the Briton's discovery, the American researcher Joseph Henry had found the same phenomena working in a different way. He is credited with the discovery of self-inductance, made during experiments with copper coils used in electromagnets. Henry found that when he switched off a current, a spark was produced. This was due to the changing electromagnetic field (formed as the current faded away), inducing a momentary voltage in the coil in the opposite direction.

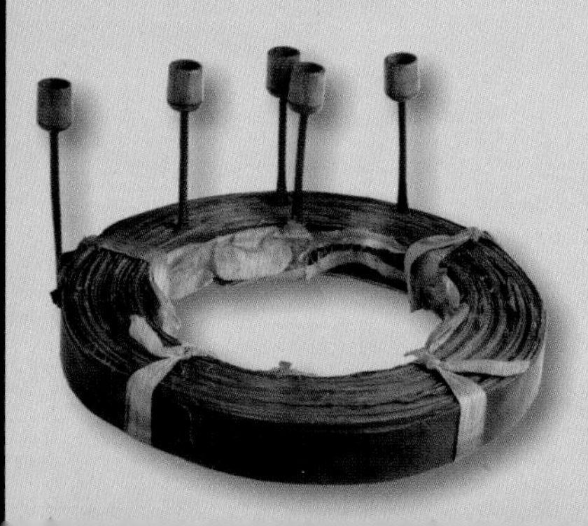

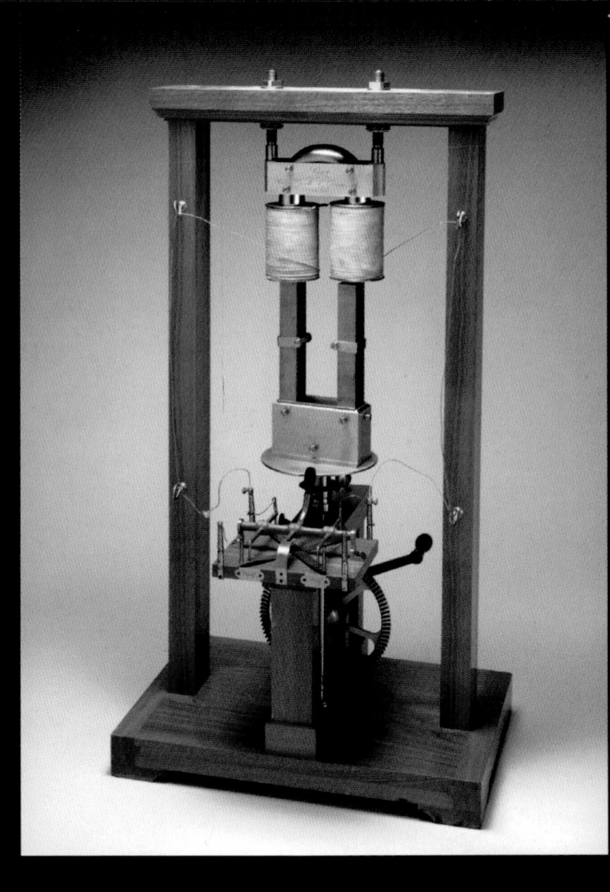

private. He investigated whether a magnet could stem the flow of electricity. The answer was no. Neither did magnetism affect the path of light.

Once Davy had died, Faraday devoted more time to electromagnetism, and in 1831 he made his big breakthrough. He knew that passing a current through wire coiled around iron made an electromagnet. What about two coils connected by an iron ring? When one coil carried a current, Faraday detected a momentary flicker of electricity in the other. This was electromagnetic induction. Faraday showed that the current (actually the voltage, or force pushing the current) was being induced by a change in the magnetism. Rotating a magnet means the "magnetic field" is changing constantly, and a constant voltage is induced. Induction is the phenomenon behind today's electricity generators, which power the modern world.

Pixii's dynamo was the first device to generate current according to Faraday's law of induction. A permanent magnet was spun around, inducing a current in a wire. The device was invented by Frenchman Hippolyte Pixii in 1832.

39 The Doppler Effect

WE'VE ALL HEARD IT IN ACTION AS EMERGENCY VEHICLES HURRY PAST, SIRENS BLARING. The change in pitch we typically detect is called the "Doppler effect," but it was originally proposed as a phenomenon of light.

In 1842, Austrian physicist Christian Doppler wondered how the wave nature of light might affect it as rays traveled to Earth from distant stars. Newton had proposed that color was the result of the eye picking up wavelengths of light (blue has a shorter wavelength than red). So what happens if the source of that light was moving? A ship sailing into waves meets each crest more frequently. Doppler said that the oscillations of starlight that is moving toward us would do the same—increasing the frequency and making the star look blue. A "red shift" would result when the star was moving away. Doppler suggested that the same effect worked with sound waves, and this was proved to be the case by Buys Ballot in 1845.

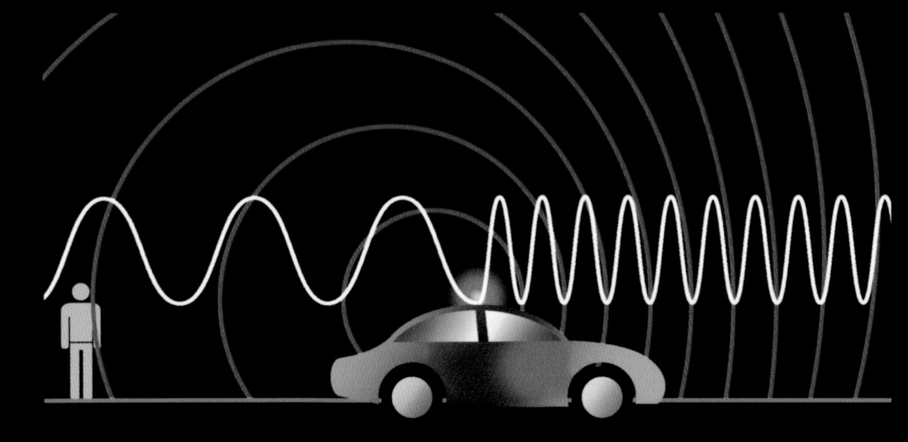

The sound wave from the approaching police siren is compressed into a high-pitched wail. As the source of the sound passes and moves away, the pitch drops as the sound wave is stretched, into a longer wavelength.

40 Thermodynamics: The First Law

JULIUS ROBERT VON MAYER'S INSPIRATION CAME FROM UNUSUAL SOURCES: The waves of stormy seas and the color of blood. However he arrived at it, his conclusion that "Energy is neither created nor destroyed" formed the first law of thermodynamics—the study of heat.

Mayer had an unhappy middle age. The deaths of two of his children affected his mental health. He spent the whole of the 1850s in an asylum.

The hero of this story is often referred to as simply Robert Mayer. As a young man he was studying to be a doctor in Tübingen, Germany, but fell foul of the authorities for getting involved in radical politics. Forced to leave in 1837, he still managed to qualify as a doctor and opted for work abroad, eventually signing on as physician aboard a ship headed for the Dutch East Indies (now known as Indonesia).

Oxygen is carried by the blood's red cells. They contain a carrier chemical called hemoglobin, which looks red because it contains iron. Hemoglobin on its own is a deep red, but when it is carrying oxygen the color becomes brighter.

Hot blooded

On the voyage he noticed two phenomena. Firstly, the waves of storm-tossed seas were warmer than calm waters. Was the motion heating the water? Secondly, the blood that gushed from wounded crewmen was brighter red when the ship was in warm climes. Bright blood contains more oxygen than dark blood. Mayer realized that the sailors used less oxygen to heat their bodies in warm conditions (the excess stayed in their blood). Mayer saw this as evidence that heat and mechanical work (objects moving) were two forms of the same thing, able to transform from one to the other. In an 1842 paper, Mayer also proposed that this energy was conserved; the total stayed constant as it switched between forms. This law of conservation of energy is a fundamental concept of thermodynamics.

41 The Mechanical Equivalent of Heat

THE JOULE

The unit of energy is the joule (J) named in honor of the English physicist. One joule equals one newton meter—the energy needed to apply a force of one newton (N) over a distance of one meter. A newton is the force needed to accelerate one kilogram to a speed of one meter per second in a second. Putting that all together: 1 J = 1 Nm = 1kgm^2/s^2.

THE SCIENTIFIC COMMUNITY DID NOT GRASP THE SIGNIFICANCE OF ROBERT MAYER'S SOMEWHAT INTELLECTUAL BREAKTHROUGH. Shortly afterward, Englishman James Joule investigated the same ideas in a more accessible fashion by performing a famous experiment.

James Prescott Joule was raised to run the family brewing business. However, he was also tutored by the great John Dalton and developed a fascination with science as a result. Initially, he investigated electricity—frequently giving the household electric shocks. His first discovery was a quantitative link between the heating of a wire and the electric current running through it.

Mechanical equivalent

Was there another way to quantify energy? Joule considered replacing the brewery's steam engines with electric motors. First he wanted to compare the efficiency of both systems and for that he needed to how much mechanical work you could expect to get from a supply of heat.

In 1843, Joule devised an apparatus to measure the mechanical equivalent of heat. A falling weight was made to spin an agitator held inside a sealed vessel filled with water. Joule's theory was that the motion energy of the falling weight, once transferred to the water, would cause it to heat up. The vessel contained a pound of water (440 g), and Joule wanted to measure how much work was needed to heat that volume by 1°F. This would give him a value for a unit of heat. His initial result was that it took 838-foot-pound force (ft·lbf). This is the energy released by dropping a one pound weight 838 feet. Later revisions and new tests using gas in place of water gave a result of 772.7 ft·lbf, much closer to the modern accepted value for this work.

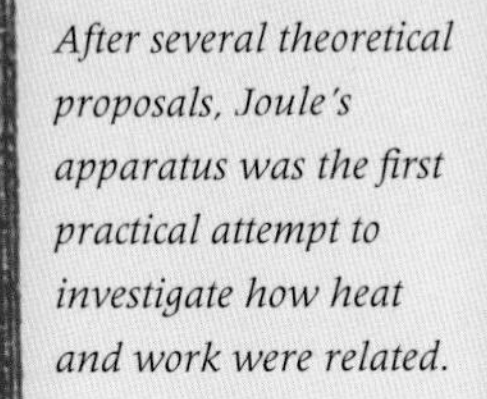

After several theoretical proposals, Joule's apparatus was the first practical attempt to investigate how heat and work were related.

42 One Energy

LIKE MANY A PHYSICIST OF THE DAY, HERMANN VON HELMHOLTZ WAS A MEDICAL DOCTOR who harbored a passion for science. His broad range of skills and interests helped reveal an underlying physical fact.

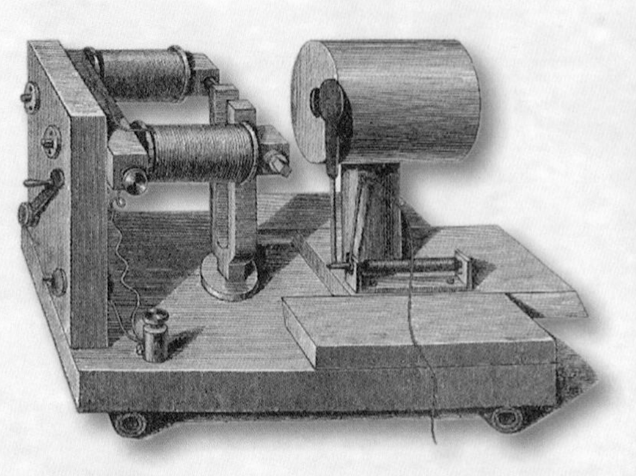

Helmholtz's resonator was a hollow cylinder that produced sound tones when struck by a tuning fork made to vibrate by an electric field. Alexander Graham Bell mistook the device for a transducer, turning electrical signals into sound—and went on to invent the telephone!

The young Hermann was coerced into medicine by his father, because there was financial support for medical students in Germany at the time. (Studying pure sciences was the preserve of the independently wealthy.) Helmholtz is included in the history of physics because he was the first to expand the scope of our understanding of energy. He took an interest in the physics of sound, light, and optics (inventing the ophthalmoscope for looking inside the eye), the electrical impulses in nerves, and the force fields around electromagnets. Helmholtz realized that all these phenomena were manifestations of a single energy, frequently transformed but always conserved. He published his findings in 1847 alongside the first mathematical proof of the concept.

43 Absolute Temperature

WHAT WAS ENERGY? JAMES JOULE HAD SUGGESTED THAT HEAT energy was in fact the motion of atoms and molecules. That prompted one Scottish scientist to ponder what happened if they stopped moving.

From the time of Black and Lavoisier, heat had been imagined as a fluid that moved between objects. James Joule's proposal seemed to match the evidence better. It explained why hot gases exerted a higher pressure—their molecules moved around more quickly hitting things more frequently. Changes of state—going from solid to liquid to gas—can be understood as molecules begin to move independently of each other: In solids they are bonded together; as liquid, the molecules begin to break free and flow around

William Thomson, now ennobled to the 1st Baron Kelvin, gives a lecture at the University of Glasgow, Scotland, where he was professor of natural philosophy for 53 years.

each other; and then in gas form, they are all free to move independently. The Scottish physicist William Thomson (later elevated to Lord Kelvin) described the energy of motion as "kinetic"—and Joule's ideas about moving molecules became kinetic theory.

Temperature is a measure of the average kinetic energy of the molecules in a substance. In 1848, Thomson proposed an absolute temperature scale, now known as the Kelvin scale (K). This used the same degree unit as defined by Celsius in 1742. However, instead of being the freezing point of water, 0 K marked the point when the atoms of a substance had zero kinetic energy. Thomson calculated that this "absolute zero" was the equivalent of –273.16 °C (–459.67 °F), the lowest temperature imaginable.

44 Working at Light Speed

GALILEO WAS THE FIRST TO TRY TO MEASURE THE SPEED OF LIGHT, USING LANTERNS POSITIONED FAR APART. He failed. To his human eye the light appeared to move instantaneously. But science would find a way to measure beyond human perceptions.

The first measurement of the speed of light was made by the Danish astronomer, Ole Rømer, in 1676. He calculated the orbit of Io, the first moon of Jupiter (one of four discovered by Galileo in 1609), pinpointing when the moon would be visible from Earth. He then used the time lag between its theoretical appearance and actual one to measure how long the light had taken to travel between Io and him. His method was sound but the distance measurements he relied on were not, so his figure was out by about 25 percent. In 1849, a better method was proposed.

A contemporary illustration of Fizeau's light-speed equipment. The light from the source was reflected off an angled mirror toward the distant mirror (left). An observer aligned the beam at the reflecting end, and the returning beam passed through the angled mirror to a second observer (right).

Mirrors and cogs

Frenchman Hippolyte Fizeau built an apparatus to measure the speed of a light beam he controlled. He shone a powerful light onto a mirror about 8 km (5 miles) away, passing it through the gaps between the teeth of a spinning cog on the way. The cog's teeth never turned fast enough to block the light completely, but at a certain speed, the light reflecting back was seen to dim. This was because the returning ray was blocked by a cog tooth swinging in front of it. Fizeau could then calculate the "time of flight" of the light beam from the spin speed of the cog. His result was 313,300 km/s (194,675.6 miles per second)—still 4 percent out. Today's figure is 299,792 km/s (186,282 miles per second), and as we shall see later this is the speed limit of the Universe.

45 Spectroscopy: Essential Information

AS LENS TECHNOLOGY GREW MORE ADVANCED, SCIENTISTS WERE ABLE TO OBSERVE AND MANIPULATE light with ever greater accuracy. Where Newton had seen a full rainbow of color in sunlight, new optical instruments showed gaps in the spectrum. What did they mean?

On hearing of Ole Rømer's measurement of the speed of light by observation of a Jovian moon, Isaac Newton's first response was to ask what color the light was. When he heard it was white, he was satisfied that light travels at the same speed, whatever its color. However, the colors from stars (and anything else that glows hot) did have a tale to tell—and gave an early peek inside the atom itself.

Above: Robert Bunsen and Gustav Kirchhoff pictured during their collaboration in Heidelberg in the 1850s.

Left: Bunsen developed a clean gas burner for performing flame tests, a device that still bears his name and is used in laboratories the world over.

Dark and light

In 1814, a Bavarian lensmaker called Joseph von Fraunhofer perfected a form of glass that did not suffer from chromatic aberration—coloring effects that obscured objects and often gave a false image of them. He built a device called a spectrometer (see box below) to show off the colors in a source of light in exquisite detail. He found that when he looked at the light coming from the Sun, there were dark gaps between the otherwise seamless blend of colors.

Named Fraunhofer lines, these anomalies were left unexplained for 40 years. The explanation came in 1859 from a pair of chemists in Heidelberg who were using a spectrometer to study the color of flames. Robert Bunsen and Gustav Kirchhoff split the flaming light into a series of discrete color lines—the opposite of what Fraunhofer had seen. Each of these "emission spectra" was specific to

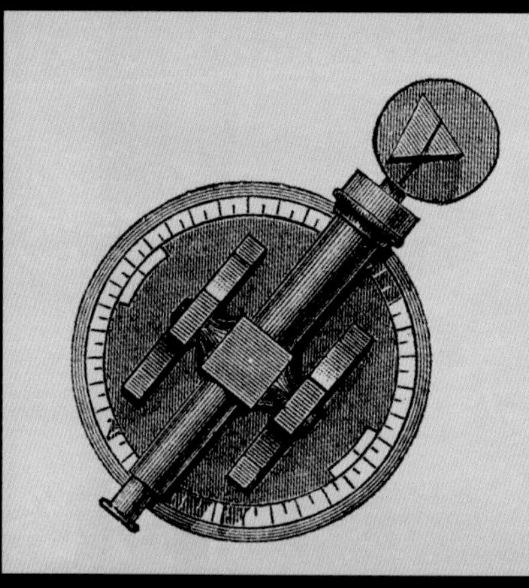

THE SPECTROMETER

Joseph von Fraunhofer's spectrometer was a prism combined with a telescope. White light, split into its constituent spectrum of colors by the refined glass prism, was then magnified in crystal clarity for the view by the telescope. A second telescope could be used to focus a beam from the source of light onto the prism.

the element being burned. The inference was that the lines seen in sunlight were an "absorption spectrum" where a certain element, this time cold, was absorbing the same telltale colors.

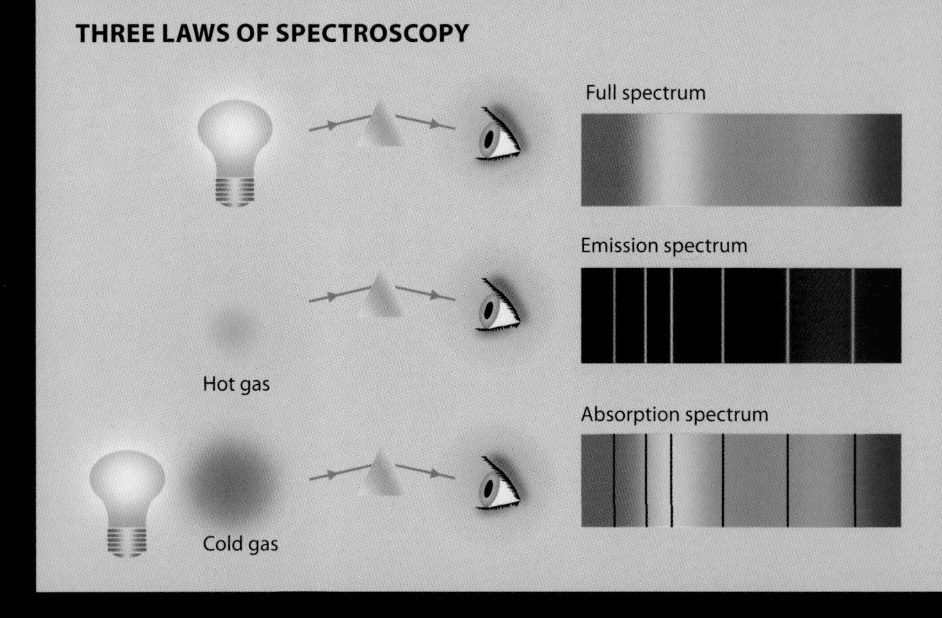

Laws of spectroscopy

Kirchhoff encapsulated the discovery with three laws: 1) Hot solids produce a full spectrum of colors (white light); 2) Hot gas glows with a specific emission spectrum; 3) Cold gas absorbs specific colors from light shining through it, leaving Fraunhofer lines in the full spectrum.

It appeared that every element took in and gave out energy as a unique set of colors, or wavelengths, of lights. Why was it that atoms dealt only in certain wavelengths? It would require quantum physics to find the answer.

46 Maxwell's Equations

JAMES CLERK MAXWELL WAS THE SUCCESSOR TO MICHAEL FARADAY, CONVERTING A LARGELY PICTORIAL UNDERSTANDING OF ELECTROMAGNETISM into four mathematical equations.

Michael Faraday explained the attractions and repulsions of magnets and electrical charges in terms of "lines of force," a graphical device still employed in school rooms to this day. The lines carried arrows showing how they would interact with one another, and together they made up a force field around an object. Faraday did no research in later life, and by the mid-1860s, Maxwell, a Scottish mathematician, had become the leading figure in electromagnetism.

Over the previous ten years, Maxwell had been studying how force fields changed and found that the rate of change was the speed of light! Here was a clear link between the phenomena of light waves, electric currents, and magnetic forces, and Maxwell took up the challenge of unifying them. The results were the Maxwell equations, published in 1865, a set of mathematical tools that can calculate all the different variables at work in an electromagnetic field. Faraday had proposed that gravity had a field as well. Who would figure out those field equations? Step forward Albert Einstein.

A year after graduating James Maxwell had proved that white light could be formed from combined red, blue, and green beams. He became full professor at the age of 25.

47 Going from Hot to Cold

IN 1850, RUDOLF CLAUSIUS HAD PROPOSED AN EARLY VERSION OF THE SECOND LAW OF THERMODYNAMICS: "Heat cannot of itself pass from a colder to a hotter body." It appeared that thermodynamics had a direction, and Clausius proposed a reason why.

Clausius's concept of entropy explains why wires always end up tangled when left to their own devices. However neat you make them, they will always become more disordered.

Clausius's work in thermodynamics was built on the obvious flaw in Carnot's theoretical heat engine. It relied on all heat being converted to work and then all that energy being converted back into useful heat. Even discounting imperfections in the mechanism, Clausius realized this was never possible. There would always be less heat left over by each iteration of the cycle—and the mechanism would grind to a halt.

The heat energy did not disappear, it had just left the system, heating the surroundings. To keep the engine going, therefore, energy must always be added. In 1865, Clausius described the decrease in available energy as an increase in another new concept: Entropy, a measure of disorder. That allows us to restate the second law of thermodynamics: The Universe is moving toward equilibrium, a state of maximum entropy, where energy is evenly distributed and no longer flows from one object to the next.

Clausius looked old before his time. His wife had died young, leaving him six children to raise alone.

48 Electrifying Gases

WHEN VACUUM PUMPS BECAME GOOD ENOUGH, SCIENTISTS FOUND THAT ELECTRIFYING TUBES OF HIGHLY DIFFUSE GASES made them glow. But this was no ordinary light. Was it really light at all?

In the 1850s, the German instrument-maker Heinrich Geissler invented what became known as the Geissler tube. Having sucked as much gas out of a glass vessel as possible, he ran a current through the near vacuum. An eerie glow formed—and the color varied depending on what gas was in the tube.

What Geissler had invented was an early version of the gas-discharge lamp. Later the same concept would be used in so-called neon lights, and today's low-energy light bulbs are the latest version of the Geissler tube. As a skilled glassblower, Geissler started producing tubes in a range of shapes, sizes, and colors. To most they were little more than scientific curiosities.

An illustrated montage from the turn of the 20th century shows, from right to left: Geissler with his pretty tubes, Plücker using a coiled electromagnet to investigate the material properties of the glow; and finally Crookes shows off his cathode-ray tube with a Maltese cross on the end. The ray was powerful enough to cast a shadow of the cross on a screen.

Then Geissler's colleague at the university of Bonn, Julius Plücker, discovered that the soft light that filled the tube could be deflected by magnets. This was no ordinary light at all.

By the 1870s, pumps had got a lot better, and English physicist William Crookes made a Geissler tube with 10,000 times less gas inside and with a much larger current applied to it. It produced a different effect: The current produced an invisible ray, starting at the negative electrode—or cathode. By the time this ray reached the anode (positive terminal), an eerie glow was beginning to form. But the ray did not stop there but carried on in the same direction causing a fluorescent coating at the end of the "cathode-ray tube" to glow. As Plücker had found, the cathode ray was deflected by a magnetic field. It even made a paddle wheel inside the tube spin. What was in this mysterious beam?

49 Boltzmann's Equation

IN THE 1870S, LUDWIG BOLTZMANN MANAGED WHAT FEW ACHIEVE: HE FOUNDED A FIELD OF PHYSICS. HIS CREATION WAS STATISTICAL MECHANICS, which uses mathematics to model the motion of invisible atoms and molecules. His system worked, but was it correct?

$$\frac{\partial f}{\partial t} = \left(\frac{\partial f}{\partial t}\right)_{\text{force}} + \left(\frac{\partial f}{\partial t}\right)_{\text{diff}} + \left(\frac{\partial f}{\partial t}\right)_{\text{col}}$$

Statistical mechanics uses mathematics to explain the large-scale, visible properties of a substance in terms of the motion, mass, and other attributes of the invisible atoms and molecules inside it. Boltzmann spent his life in Austria and formulated the techniques of statistical mechanics while professor of mathematics at Vienna university.

Among the math tools was the Boltzmann equation, which describes the distribution of particles in a gas or liquid. It can be used to calculate how the particles collide with each other and how physical attributes, like heat or charge, flow through a substance. This in turn allows us to predict its ability to conduct heat or electricity—even how runny it is. However, this and Boltzmann's other bits of math presupposed that atoms and molecules existed, something that was far from certain even by the 1870s. Boltzmann's work was shunned by many of the top scientists of the day. Of course, atomic theory would one day be proven to be true, yet Boltzmann's techniques worked so well that even doubters, like Max Planck, could not resist using them—to great effect.

50 Tesla: An Alternating Character

A picture of Tesla in his late thirties, then based in New York City and becoming a leading figure in the new industry of electrification.

NIKOLA TESLA COMBINED SHOWMANSHIP AND INVENTIVENESS TO BECOME THE ARCHETYPAL 19TH-CENTURY MAD SCIENTIST. His successes were enormous, forming the basis of today's electrical generation and supply system, but he also had several high-profile failures. While others grew fabulously wealthy from his inventions, Tesla would die in poverty.

Tesla helped to make the modern world. The generators in our power plants, the substations that feed power to our communities, and even the electric motors in the very latest electric cars all owe a debt to this mysterious genius.

Tesla made these contributions in the United States at the end of the 19th century, when the country was emerging as an unrivaled industrial superpower. After a short stint working for Thomas Edison's European operation in France, Tesla moved to New York in 1884 and began to work for the man himself. In a few years he had completely redesigned Edison's inefficient direct current (DC) generators.

The Pittsburgh laboratory where Nikola Tesla and George Westinghouse developed the apparatus for an alternating current electricity system.

DIRECT AND ALTERNATING

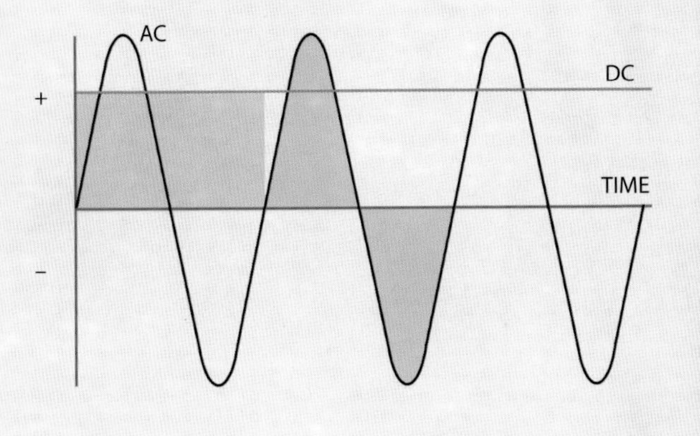

An alternating current (AC, the black curve) is in constant flux, as the force, or voltage, pushing it along oscillates first one way (+) then the next (–). AC carries energy just as effectively as direct current (DC). The blue line above shows the DC equivalent to the AC. The shaded areas represent the same amount of power in either system. Modern electronics need current to move in just one direction, and so the AC power supply is rectified into DC at the plug.

THE TESLA COIL

Looking like a metallic toadstool that spits lightning, a Tesla coil is essential equipment for any mad scientist. Invented in 1891 by Tesla, the coil is a transformer that can induce very large voltages. The primary and secondary coils are not connected by a metal core as in regular transformers. Instead an enormous charge builds up in the primary coil. When this is allowed to dissipate rapidly, it induces an immense surge of current in the secondary coil (topped with the donut shape)—which duly produces an impressive display of sparks. Small Tesla coils were used in early radios. Large ones are still used today to simulate lightning strikes or electrical accidents.

Edison was not uninterested in Tesla's ideas for an alternating current (AC) system, and a clash over money (or the lack of it) prompted Tesla to start up in business on his own.

An alternative system

Alternating current is what is produced by the simplest electricity generators. As the poles of the magnet spin around, they induce a current first in one direction, then the other. In 1887, Tesla invented the AC induction motor. This has a magnetic rotor that is made to spin by the constantly changing magnetic field created by AC.

The following year, Tesla was employed as a consultant by George Westinghouse, Edison's great rival. The pair worked together rolling out electricity supply services, this time provided as AC. Generating AC is more efficient than DC. Also with Tesla's help, the Westinghouse system used transformers to manipulate the voltage of the supply. A transformer is a device that uses induction to change the voltage of an electric current. The device has two coils of wire both connected to the same core, generally a lump of iron. An AC current in the primary coil induces a current in the second coil. If the primary coil has fewer turns of wire than the secondary one, the transformer will give out a current with a higher voltage than it receives. The opposite situation results in the voltage stepping down. Step-up transformers are used to create high-voltage currents for transmission over long distances. This is the most efficient way of doing it, minimizing heat losses. (Boosting the voltage of DC is a much more complex task.) However, high voltages will make your household appliances literally explode, so the supply is stepped down at substations before it reaches your house. Thanks to Tesla, the world's electricity comes in AC, even today.

51 Mach Goes Supersonic

SOUND IS A WAVE OF PRESSURE, AND LIKE ANY WAVE IT HAS A FINITE SPEED. The name of German physicist Ernst Mach has become intimately associated with what happens when we break that speed limit.

The moment a fighter jet becomes transonic (breaking the sound barrier) is captured. The cloudy vapor cone is produced by the pressure wave surrounding the aircraft.

Sound is a wave moving through a medium, creating a series of compressions and rarefactions. Our ear is tuned to detect the vibrations in the air and process them into something we can hear. The speed at which a sound wave moves depends on the medium through which it travels. The speed of sound for a particular medium depends on a number of variables, but generally it moves faster through materials that are harder to compress. For example, sound travels through water more than four times faster than in air, and even faster still through rock.

$$M = \frac{v}{v_{sound}}$$

In 1887, after a career of studying the behavior of light waves, Ernst Mach turned his attention to sound. At the time there was no well-formed idea that sound waves were structurally different to those of light, and Mach was interested in what happened when an object moved through a medium faster than waves could. At the time the only things that could move at "supersonic" speeds were bullets, and Mach succeeded in photographing one at full speed a year later. As he had predicted, a cone-shaped shock wave surrounded the bullet. This shock wave is heard as the crack of the gunshot (or the sonic boom of a fast jet today). The ratio of actual speed and speed of sound is now known as the Mach number in honor of the German: Mach 1 is the speed of sound; Mach 2 is twice the speed of sound, and so on.

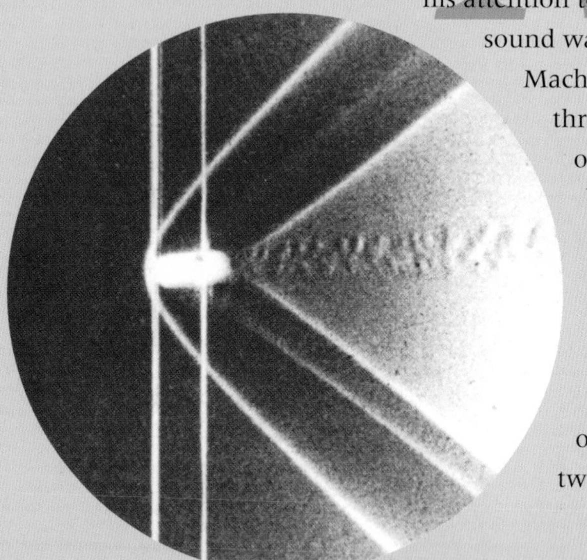

Ernst Mach's photograph from 1888 shows the cone-shaped shock wave around a bullet traveling at supersonic speeds.

52 Looking for Ether

IF SOUND WAS A WAVE RIPPLING THROUGH A SUBSTANCE, SURELY LIGHT WAS SOMETHING SIMILAR? Even in the 1880s, it was widely believed that light was a wave in an invisible and undetectectable medium called ether.

We can see light coming from the stars and so it was believed that ether must fill the space between us and them—in other words, fill the entire Universe. Despite the advances since the days of Huygens and Young, ether, Aristotle's "fifth element," was still the most compelling theory about how light traveled through the emptiness of space—and vacuums down on Earth.

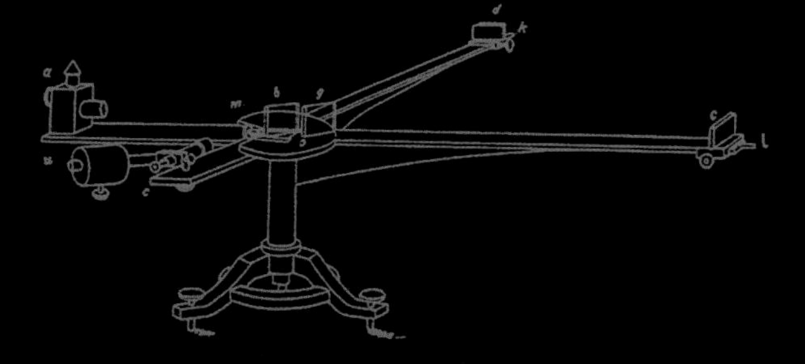

In 1881, Albert Michelson had tried to detect the ether wind using this apparatus, but found no evidence. The 1887 Michelson–Morley experiment was an attempt to refine the technique.

It was suggested that the best evidence of the existence of ether would be the drag caused by the "ether wind." As Earth moved through the ether, light traveling into the flow of ether would be slowed down—albeit very slightly. In 1887, Americans Albert Michelson and Edward Morley designed an apparatus to prove it. The device split a beam of light in two, sending them off to distant mirrors which reflected them both back to the start point. The theory went that one beam would be slowed by the ether wind, and the slight time lag would create telltale interference patterns when it met the other. The experiment was a failure, one of the most illustrious in science history, because it consigned ether to the trash can and paved the way for an alternative way of understanding the nature of light.

This is an updated model of the Michelson–Morley apparatus from 1930. It showed that the speed of light was not affected by the relative motion of Earth. The difference was that by this date Albert Einstein had explained why.

53 Waves Through Nothing

IN HIS OLD AGE, JAMES CLERK MAXWELL HAD PREDICTED THAT LIGHT WAS NOT THE ONLY TYPE OF WAVE ASSOCIATED WITH ELECTROMAGNETISM. In 1887 (it was proving quite a year for physics), a sparky young German called Hertz was about to make quite a name for himself.

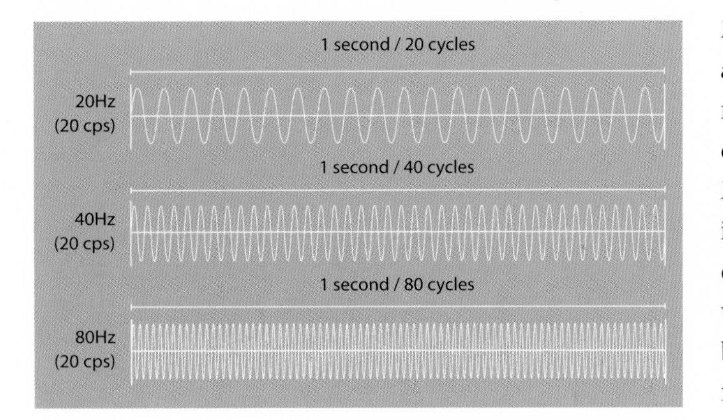

Today the term hertz is used as the unit of frequency (Hz), a measure of cycles per second (cps). From top to bottom, the diagram above shows waves of increasing frequency.

Michael Faraday had found that an electric field and its associated magnetic field were always at right angles to each other. That meant that the electromagnetic waves in light as described by Maxwell's equations did not ripple along like waves in water or like sounds. Instead, they jiggled up and down and side to side all at the same time! Such a wave was not a motion in the ether or anything else but formed wholly by constantly oscillating force fields that carried energy hither and thither with no need for a medium at all.

Different energies

Like all good theories, this one could offer a prediction. Maxwell suggested that waves that were invisible to the eye but traveled at the speed of light could be made using electricity. He was especially certain that waves that carried less energy than light would be found one day. Ten years after Maxwell's death, Heinrich Hertz, the protégé of Hermann von Helmholtz, built a device to detect these invisible waves.

Send and receive

Hertz's apparatus was designed to send sparks across a gap between two

Hertz's spark gap oscillator provided the first evidence of electromagnetic waves that you couldn't see. They were later named radio waves, derived from the term "radiation."

brass balls. The theory was that as well as producing a visible spark of light, the device would give out invisible electromagnetic radiation. To detect this, Hertz built a very simple receiver, which was just a loop of wire that also had a small gap in it.

This was placed several yards away from the spark gap, much too far away for the electrical spark to jump across to it. However, Hertz saw that a small spark did move across the gap in the receiver. This showed that invisible waves were radiating out from the device and inducing a small current in the receiving ring. The mysterious phenomena are now known to be "radio waves," which contain a good deal less energy than light: Hertz had sent the first radio signal. The next question was how many other forms of electromagnetic radiation were there to discover.

54 X: The Unknown Ray

THE DISCOVERY OF ANOTHER NEW MEMBER OF THE ELECTROMAGNETIC SPECTRUM WAS DECIDEDLY MORE HAPHAZARD THAN HERTZ'S DETERMINED SEARCH. Several researchers were suggesting that invisible rays were escaping from cathode-ray tubes.

The idea of seeing the inside of the body with X rays quickly caught the public imagination.

A cathode-ray tube produces a faint, eerie glow. The light passes through the glass, but the invisible cathode ray itself was trapped inside. In the late 1880s, Philipp Lenard modified Crookes' design of a cathode-ray tube, adding an aluminum "window" to the glass. The metal component was sturdy enough to withstand the air pressure pushing in from outside, but also thin enough to allow the elusive cathode rays to leave the tube.

German physicist Wilhelm Röntgen is thought to have been using a Lenard window when he found a mystery beam, which he recorded with a quizzical "X." The delicate Lenard window was normally protected by thick paper when not in use. While setting up his tube in 1895—it was switched on but still covered—Röntgen saw a glow on a photosensitive screen lying on his workbench. His conclusion was a new form of electromagnetic wave was the cause. With a name derived from that initial X—Röntgen found that X rays could pass through some solid objects. He used X rays to photograph the bones in his wife's hand, who was not thrilled with the result, saying: "I have seen my own death."

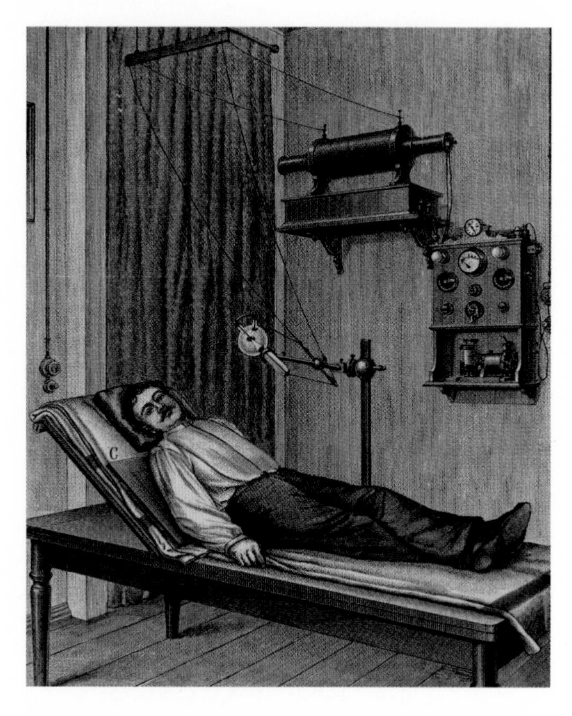

A year after they were discovered, X rays were already being used in medical imaging. Here a cathode-ray tube is shining X rays onto the patient's chest, forming an image on the photographic plate behind the poor fellow.

THE SUBATOMIC AGE

55 Radioactivity

RECHERCHES

SUR UNE

PROPRIÉTÉ NOUVELLE DE LA MATIÈRE

ACTIVITÉ RADIANTE SPONTANÉE OU RADIOACTIVITÉ DE LA MATIÈRE

PAR

M. HENRI BECQUEREL

MEMBRE DE L'ACADÉMIE DES SCIENCES

AVANT-PROPOS

Les premières expériences relatives à la découverte et à l'étude des phénomènes dont il va être question ont été publiées dans les *Comptes rendus des Séances de l'Académie des Sciences* il y a sept ans. Le détail des recherches poursuivies en 1896 et 1897 devait faire à cette époque l'objet d'un premier Mémoire. Le désir d'élucider quelques expériences contradictoires, et l'obligation de ne pas abandonner un autre travail en cours d'exécution, ne me laissèrent pas le loisir de faire cette publication.

En 1898, les travaux que M. et Mme Curie entreprirent

Becquerel's work on radioactivity was published in 1903 as Recherches sur une propriété nouvelle de la matière, *or "*Research on a New Property of Matter.*"*

CATHODE–RAY TUBES ARE NOT THE ONLY THINGS THAT GLOW IN THE DARK. SOME MINERALS DO IT TOO, OFTEN DULL-LOOKING CRYSTALS BECOME FABULOUS GEMS when the lights go out. A French physics professor wondered if these glowing oddities also give out invisible rays like the ones just discovered by Röntgen. What he found was radiation of a different sort.

The researcher was Henri Becquerel, already a leading figure in French academia. His experiment was to recreate the conditions of Röntgen's accidental discovery but replace the cathode-ray tube with samples of phosphorescent minerals. In this way he hoped to show that the minerals were also a source of X rays or another form of electromagnetic radiation. Although his methods were well thought through, the discovery they revealed was just as much a happy accident as that of Röntgen's.

A false-color photograph shows Henri Becquerel in his laboratory.

Not the foggiest

Röntgen's source of X rays had been shrouded in an opaque covering, thus proving that the observed glows were not the result of light shining from the cathode-ray tube. In the same way, Becquerel covered his detector (in this case a photosensitive plate) in a similar dark paper to ensure no light could get in. He then placed on top samples of phosphorescent minerals and artificially produced powders that were known to glow in the dark. (Today, we understand that these types of compounds glow because they absorb a certain wavelength of light and then gradually radiate it out again at a different wavelength, which is why they appear to glow when there is no other light source.)

Becquerel's theory predicted that if the glowing materials emitted X rays, those would pass through the paper cover and fog the photographic plate inside. He got no results.

No results, that is, until he tried with a sample of uranyl sulfate, a sandy mineral that contains uranium. This material is better known today as yellowcake, a precursor material to nuclear fuel. However, at the time uranium was regarded as a harmless heavy metal, used chiefly to give pottery and glassware a distinctive yellow–green pigment. All that was about to change.

Becquerel rays

The uranium salt did leave a foggy mark on the photographic plate. Obviously, the phosphorescent nature of the uranyl sulfate was not the cause, so the new phenomenon was dubbed "Becquerel rays." Becquerel took a lateral step and began to investigate other uranium-rich compounds, such as pitchblende, the metal's main ore. They all fogged their plates; uranium was the source of the rays. A few years later Becquerel's research student, Marie Curie, would rename the phenomenon *radioactivity.* Only certain "radioactive" elements are a source of this kind of radiation.

ALPHA

BETA

GAMMA

RADIOACTIVITY

The emissions of radioactive substances are three very different phenomena. Alpha and beta radiation are not radiation at all, but streams of particles. Alpha particles were found to have a positive electric charge and are blocked by almost all solids. Beta particles have a negative charge and are blocked by thin metal screens, suggesting they are smaller than alpha particles. Only gamma radiation is electromagnetic. It contains a lot of energy and is only blocked by thick lead shielding.

Three radiations

The question was whether Becquerel rays were radiation in the same sense as radio waves, light, heat, and X rays. In 1898, a young New Zealander, named Ernest Rutherford, came to work at Cambridge University in England. He was to be a huge figure in the field of atomic physics about to open up the inner workings of the atom. He did not know it yet, but his work on uranium at Cambridge would be the first step down that path. Rutherford found that there were two types of radiation coming from the metal, which he named alpha and beta. The passage of alpha radiation was blocked by a thin sheet of gold foil, while beta radiation passed straight through it. In 1900, Frenchman Paul Villiard found a third type of radiation coming from radioactive metal (the newly discovered radium) which had more penetrating power than either of those observed by Rutherford. In keeping with the convention, this was named gamma radiation. In the years to come these radioactive emissions would be at the heart of research into atomic structure. And as that era dawned, Becquerel made another discovery about his rays—they burned his skin. A 1901 report into this led to the advent of radiomedicine, the therapeutic use of radioactivity.

An optimistic vision of the future shows the year 2000 from the point of view of 1900. Apart from looking remarkably like the year 1900, the fireplace does not contain burning coals but is filled with a warming glow of radioactivity.

56 The First Subatomic Particle

PHYSICISTS WERE STILL IN THE DARK ABOUT CATHODE RAYS. THEY EXISTED IN A VACUUM LIKE ELECTROMAGNETIC WAVES, BUT SHARED PROPERTIES WITH METALS AND GASES. The description used was "radiant matter." In 1899, an English scientist proved they were particles—but smaller than even atoms!

John Joseph "J.J." Thomson is the man responsible for discovering these tiny entities. His discovery came about when he reran an experiment by Heinrich Hertz. Like Hertz, Thomson wanted to test whether cathode rays could be deflected by an electric field. Hertz's research had indicated no, but when Thomson repeated the experiment he got the opposite result. (Thomson's tube had less gas in it; the excess of gas left in Hertz's tube was becoming charged by the electric field, thus negating any effects it might have on the passing rays.)

DETECTING ELECTRONS

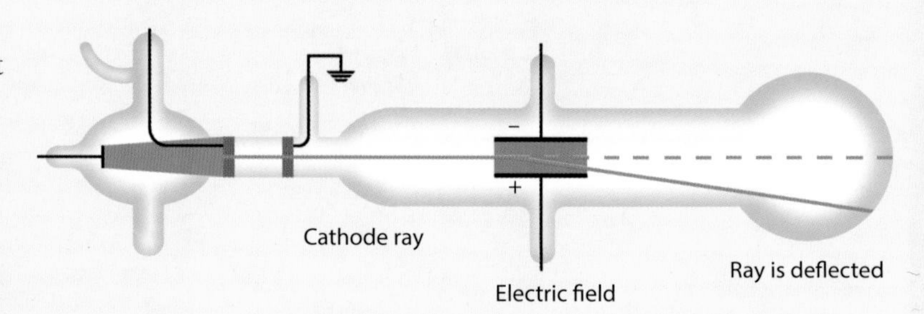

Thomson's apparatus ran a cathode ray between two charged plates which had an electric field running between them.

Thomson's results showed that the ray swung toward the positively charged side of the field. In electromagnetism, opposites attract, and so such a deflection means cathode rays have a negative charge. Light has no charge, so the beam must be invisible particles. He then compared the effects of a magnetic field with those of the electric one, to calculate the speed and charge of the beam. The astonishing result was that the particles were 1,800 times lighter than hydrogen atoms, the lightest of all atoms! The term *electron* had been coined a few years before as a theoretical charge carrier for electricity. Thomson adopted this name and proposed that the electric field of the Crookes tube was ripping electrons off atoms: That made them the first *subatomic* particles to be found. Were there others?

J.J. Thomson pictured surrounded by his state-of-the-art 19th-century technology,

57 Planck's Constant

IT HAD BEEN KNOWN SINCE ANTIQUITY THAT HOT THINGS GLOWED, AND THEIR COLOR WAS AN INDICATOR OF HOW HOT THEY WERE. Attempts to understand invisible thermal rays on the same terms sharpened a whole new cutting edge for physics.

Max Planck is the founding father of quantum physics, although he only introduced the concept of quanta to make the complex math work.

White-hot metal is hotter than red hot—but the hottest color of all is blue. Here was a clear link between the frequency (or wavelengths per second) of light, which we see as color, and the energy in the wave. But no one knew if thermal radiation—the heat we feel on our skin—varied like this.

In 1859, Gustav Kirchhoff's work on spectroscopy prompted him to phrase this question in terms of how an ideal "black body" radiated energy. (It was termed "black" because it absorbs all the radiation that hits it.)

In the 1890s, Max Planck had a go at finding an answer. He devised mathematical "springs" that took in and gave out radiation. Planck used them to model thermal equilibrium in a black body, that is when it radiates energy at the same rate as it absorbs it. He hoped to match the energy in heat waves to temperature.

At first Planck's math did not work, but in 1900 he applied Boltzmann's statistical mechanics and had more success. In the new math, Planck represented energy as quanta, or tiny, indivisible units. It worked, and showed $E=hv$: The energy (E) of a electromagnetic wave is proportional to its frequency (v). The constant of proportionality (h) is now called Planck's constant. This tiny number showed that the energy in particles could not exist in any quantity, but had to be multiples of quanta. Quantum physics had arrived.

Max Planck's other contribution to quantum physics are Planck units. When expressed in these units, universal constants—things like the speed of light and big G—all equal 1. As this table shows, the Planck units are minute quantities: In fact, they are the smallest possible magnitudes in the Universe.

Quantity	Expression	Metric value	Name
Length (L)	$l_P = \sqrt{\frac{\hbar G}{c^3}}$	1.616×10^{-35} m	Planck length
Mass (M)	$m_P = \sqrt{\frac{\hbar c}{G}}$	2.176×10^{-8} kg	Planck mass
Time (T)	$t_P = \sqrt{\frac{\hbar G}{c^5}}$	3.3912×10^{-44} s	Planck time
Temperature (Θ)	$T_P = \sqrt{\frac{\hbar c^5}{G k_B{}^2}}$	1.417×10^{32} K	Planck temperature
Electric charge (Q)	$q_P = e/\sqrt{4\pi\alpha}$	5.291×10^{-18} C	Planck charge
	$q_P = e/\sqrt{\alpha}$	1.876×10^{-18} C	

58 Long-Distance Radio

GUGLIELMO MARCONI IS REMEMBERED FOR TAMING THE POWER OF RADIO WAVES with his wireless technology. Mostly businessman and engineer, Marconi's scientific contributions revolutionized the fields of telecommunications and broadcasting.

SAVING THE TITANIC

The potential of Marconi's radio technology was illustrated during the *Titanic* tragedy in 1912. Onboard were two Marconi radio operators, who sent out the distress calls. Although these were received too late to save most of the stricken passengers (the radio was turned off on the nearest ship), without the Marconi technology on board, the death toll would have been even higher. Since the disaster, ship's radio rooms have been crewed 24-hours a day.

On December 12, 1901, the first radio message to be sent across the Atlantic Ocean was received in a chilly hut on the coast of Newfoundland, Canada. The triple tones that denoted the letter *s* in Morse code were being tapped out in Poldhu, Cornwall, 3,500 km (2,200 miles) away.

The Italian physicist and engineer Marconi had proved his doubters wrong: Distance did not limit radio waves. It had been assumed that since radio waves traveled in straight lines, just like light did, they could not travel beyond the horizon. After all, by its very definition, the horizon is the limit of light—beyond that point no light reaches us and we can't see anything. How did Marconi do it?

The young Guglielmo Marconi pictured in 1896 with the radio equipment he developed in the attic of his family home.

Hertzian wave

Marconi had not excelled at school and used his family's money to pay for an education at the University of Bologna. While there in 1894, he briefly worked under Augusto Righi, who had researched Hertzian waves. These are now known as radio waves, but back then were still named for discoverer Heinrich Hertz, who had just died. Marconi focused on increasing the power and range of Hertz's spark-gap transmitter. He also improved the receiver using a coherer, a tube of metal filings which were made to cling together (or cohere) by the induced voltage. That had the effect of decreasing the electrical resistance of the tube, allowing a burst of electricity to flow, which was converted to a click by a speaker. This made it possible to send messages via the dots and dashes of Morse code, which had been developed for the telegraph. However, Marconi's equipment required no wires!

By 1895, he was sending signals over several miles. Marconi applied for more funds from the Italian government but got no interest. His work was eventually sponsored by Britain, and Marconi began to extend the range of his equipment further. In 1899, he established a radio link between England and France. Marconi was just 24 years old.

Combining the latest wireless technology with modern vehicles, Marconi fitted a radio transmitter to a steam tractor as he traveled around Wales and England testing how far he could send signals in the late 1890s.

Charged atmosphere

To extend the range of his research, Marconi then began to transmit between ships in the Atlantic. He found he could send messages beyond the horizon—and curiously the distance was greater at night. (We now know the messages were bouncing off the ionosphere, a charged layer in the high atmosphere. At night this layer rises even higher, meaning it can reflect signals over a greater surface distance.)

Progress was slow after the 1901 transatlantic signal, but Marconi developed ship radio systems, and his eponymous company was instrumental in developing broadcast technology.

BROADCASTING LEGACY

Marconi called transmitting voices via radio signals "the wireless telephone." However, others beat him to the first audio broadcasts. In 1915, he developed ways of transmitting continuous-wave sound signals using vacuum tubes. In June 1920, they were used to broadcast a performance by Australian soprano Dame Nellie Melba (right) across Europe from his English headquarters in Essex. The Marconi broadcasting operation later became the BBC, the British Broadcasting Corporation, the world's largest public-service broadcaster.

59 The Curies

THE MOST FAMOUS COUPLE IN SCIENCE, MARIE AND PIERRE CURIE ARE CHIEFLY REMEMBERED FOR THEIR WORK ON RADIOACTIVITY. However, the Curie name is also linked to other areas of physics.

In a device invented by the Curies, a high-voltage field is used to deflect the beams of alpha and beta radiation coming from a radioactive sample. The gamma rays are not deflected, while alpha radiation is positively charged and so goes in the opposite direction to the beta particles. In 1900, Henri Becquerel showed that beta radiation was in fact a stream of electrons.

Marie Curie holds a special place in science history, as the first (of very few) women to win a Nobel prize (and the only one to win two!). Nevertheless, her husband Pierre should not be overlooked. He had already earned his place in the hall of physics fame before meeting Polish émigrée Marie Skłodowska at the Sorbonne in Paris in 1894. A decade before, he and his brother Jacques discovered piezoelectricity, a phenomenon where a substance releases electricity when squeezed. However, he is best known for the Curie point, a critical temperature at which a magnet loses its magnetism.

Radioactive research

Marie and Pierre (along with Henri Becquerel) won the 1903 Nobel prize for physics for their work on radioactivity. They had found that thorium was also radioactive, and that pitchblende, a mineral rich in both uranium and thorium, was emitting more radiation than expected. The obvious reason was that there was another element in there. It took four years to refine a working sample of it from half ton of pitchblende. The sample was found to contain not one but two new elements—polonium, named for Marie's homeland, and radium. Pierre died in 1906, but in 1911 Marie won the Nobel prize for chemistry for these discoveries.

The Curies were not well funded and had a lab in a drafty shed. Pierre had back trouble, so Marie did much of the heavy work, even when it was barely above freezing inside. Marie used this as an opportunity to see the effect of the cold on radioactive emissions—they were independent of temperature.

60 Einstein's Annus Mirabilis

THE YEAR 1905 IS KNOWN IN PHYSICS CIRCLES AS THE *ANNUS MIRABILIS*, OR "MIRACLE YEAR." THE MAN PERFORMING THE MIRACLES—there were four in all—was none other than Albert Einstein. The young German genius changed the way we understand matter, energy, space, and even time.

MOON FOUNTAINS

In the 1950s, space scientists predicted the photoelectric effect would do odd things to Moon dust. The tiny specks would be charged by sunlight and repel each other so they floated off the ground: Surveyor moon probes (below) confirmed that "moon fountains" existed.

First Einstein tackled the photoelectric effect: Shining light onto electrodes increases the electric current running through them. Einstein explained this effect by saying light was a stream of particles—as well as being a wave. He said each particle (later named photons) carries a quantum of energy. When photons hit a conductor, they transfer energy to it making electrons flow into a current. In reverse, objects emit energy as photons. Einstein also explained Brownian motion using kinetic theory—in other words, how atoms moved. Later he turned to the relationship between energy (E) and mass (m), coming up with the legendary equation $E=mc^2$. The value c was the speed of light, a very large number. So energy was equivalent to a mass multiplied by the square of a very large number, which meant even tiny masses contain enormous energies. The speed of light was also integral to a fourth breakthrough that year, one that made Einstein world famous: Special relativity.

History reflects that Albert Einstein was the most influential physicist in modern times. However, in 1905, the jaw-dropping theories of this unknown patent clerk were too much for many physicists to accept at first.

61 Special Relativity

ALBERT EINSTEIN'S SPECIAL THEORY OF RELATIVITY IS COUNTER INTUITIVE—SOME OF IT SEEMS TO BE COMPLETE NONSENSE. It tackled physics at high speed, the fastest of all, in fact. What was the Universe like when you were traveling at the speed of light?

Despite its mercurial revelation in 1905, Albert Einstein's most famous theory—relativity—took years to develop. The complete "general" theory was presented in 1916. The 1905 paper outlined the effects of speeds at or near to the speed of light (and has become known as the "special" theory). Legend has it that the theory had been born a decade before, when the still-teenaged Einstein asked himself: "What would you see if you sat on a beam of light?" The answer resulted in a new and very strange way of understanding the way the Universe works.

Wave direction

Light was now firmly established as a wave, and Maxwell's electromagnetic equations provided a mechanism by which these waves could travel without a medium. So how did such a wave behave? Was it possible to break the "light barrier," just as Mach had showed you could do with sound?

Einstein's teenage thought experiment was a good place to start. The speed of light is a constant. As you sit on your personal photon, whizzing along at that speed, you would think that any other photons coming from sources behind you could never catch up. So when you turned around, the Universe would be perpetually dark—no light could reach your eyes. And what about the other way? The light shining in the opposite direction to your photon would pass you at a relative speed of twice the speed of light.

In the human world, all these explanations seem perfectly reasonable, but no one had been able to detect a variation in the speed of light. Einstein's 1905 special theory said that it was impossible. The speed of light was always the same. The relative speeds of its source and the observer had no effect, and so everything would just look normal from your perch atop a light beam.

Space and time

How does motion have no impact on the speed of light? To explain, physicists such as Hermann Minkowski recast our understanding of space and time into an interconnected whole, or space-time. As an object moves past an observer, space-time from their point of view contracts, making the object shrink in the plane in which it is moving. Its mass also increases, and as a result more energy is needed to move it faster. If it were to move at the speed of light, the object would have to become infinitely massive and require infinite energy to achieve that speed—obviously impossible. One of the take-home facts of special relativity is that no mass can travel at the speed of light, but the massless photons that make up light itself can. Traveling close to light speed also makes a mass appear to move more slowly through time relative to a stationary observer.

These changes in mass, space, and time are imperceptible to us but they ensure that the speed of light is a constant relative to any observer, no matter their speed.

Next time you are in traffic think about the lights coming from the other cars. They obey a speed limit just like you.

62 A Positive Discovery

ATOMS DO NOT HAVE AN OVERALL CHARGE, SO IF ELECTRONS CAME FROM INSIDE ATOMS, THERE MUST BE SOMETHING POSITIVELY CHARGED in there as well. In 1909, Ernest Rutherford led a team to find out what it was and gave rise to a term that still resonates today—nuclear physics.

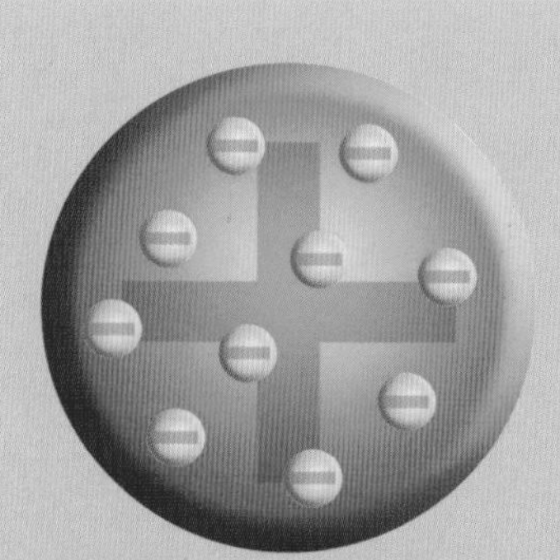

PLUM-PUDDING ATOM

J.J. Thomson had suggested that electrons were scattered through an atom like plums in a pudding. They made up just a tiny part of the atom's mass and so the "pudding" part of the atom must make up the bulk of the mass, but carry a charge that was equal and opposite to the combined charge of the electrons. The "plum-pudding model" was basically the first guess at how an atom was structured—a good start but soon to be superseded.

By this time, Ernest Rutherford was already a Nobel prize winner. He had been awarded the accolade following work in Montreal's McGill University. In 1901, he and his assistant Frederick Soddy noticed that thorium gave out a gas as well as radioactivity, and chemical analysis showed that radium had formed where the thorium had been. They had discovered the mechanism behind radioactivity—atoms of certain elements were unstable and collapsed into atoms of another element,

Hans Geiger and Ernest Rutherford (right) pose with the detector screen that shows the evidence of reflected alpha particles—the first evidence of the atomic nucleus.

emitting charged particles in the process. The atomic model of the day, the plum-pudding model, (see box, opposite) made allowances for negatively charged electrons (beta radiation) to leave the atom. But it did not provide a reason for the positively charged alpha particles.

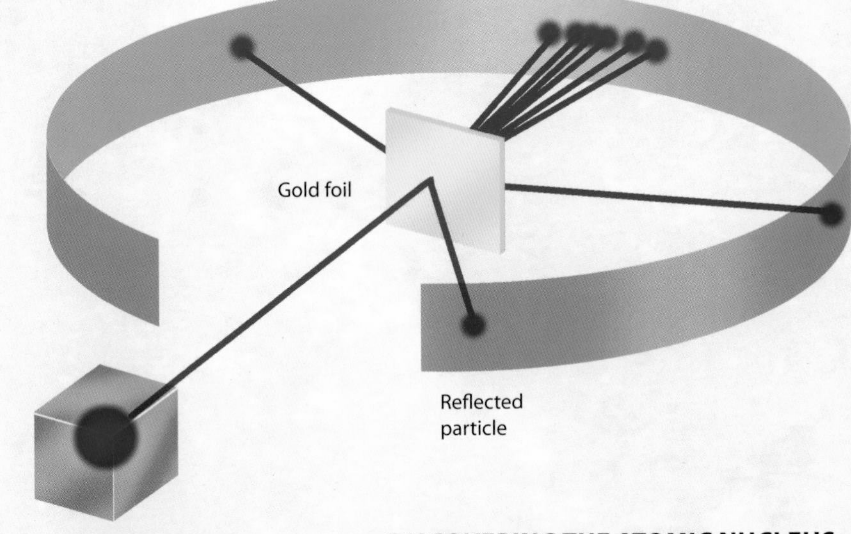

DISCOVERING THE ATOMIC NUCLEUS

The apparatus used in the so-called Geiger–Marsden experiment in 1909.

Proof of the pudding

By 1909, Rutherford was back in England, working at Manchester University, the city where John Dalton had first proposed the existence of atoms a century before. Rutherford recruited two researchers, Hans Geiger (he of the radioactivity counter) and Edward Marsden, to use alpha particles to probe the structure of the atom, which he believed to be more complex than Thomson had proposed. A plum-pudding atomic structure demanded that the electrons be spread perfectly evenly. Rutherford wanted to test this by firing positively charged alpha particles through gold foil. If the gold atoms were perfectly spaced plum puddings, the particles would go straight through without deviation. And that is more or less what happened.

Bouncing back

Not to be foiled by the foil, Rutherford suggested that before abandoning the project, his team place detector screens all around the foil, just to be certain nothing was bouncing back. This new setup did indeed show that a tiny number of alpha particles were ricocheting off the foil. Rutherford is reported to have done the *haka*, a Maori war dance, when he heard the news. To him the result was "as if you fired a 15-inch shell at a tissue paper and it came back and hit you."

Rutherford interpreted the results as the alpha particle being repulsed by the positive part of the atom. This happened only occasionally because the positive region was a minute core, or nucleus. The electrons orbited on the edge of the atom like planets, held in place by attraction to the nucleus. But the atom was largely empty space, hence why most alpha particles sailed on through. This nuclear view of the atom prevails today, although quantum physics would soon inject a lot of complexity.

63 Units of Charge

MANY PHYSICISTS COULD NOT ACCEPT THAT LIGHT AND ELECTRICITY WERE MADE OF PARTICLES. THE EVIDENCE FOR THEM BEING CONTINUOUS WAVES was just too strong. In 1909, two American colleagues set out to prove the particle idea wrong. However, they managed to do the opposite.

Professor Robert Millikan led the assault, ably assisted by Harvey Fletcher. The pair had devised a way of measuring electrical charge. Electromagnetism had been seen to behave like a wave, oscillating between two extremes, and so Millikan and Fletcher expected to find that electrical charge could take any value between those two points.

To prove it, they performed the now famous oil-drop experiment. This involved spraying fine droplets of oil into a strong electric field set up between two plates. As the scientists watched through a microscope, the droplets were allowed to fall under gravity to the bottom of the plate. Then the electricity was turned on. Some of the droplets were then charged by friction as they were sprayed out, and so these were pushed back into the air by the electric field.

The researchers then selected a suitable droplet and calculated its size and weight from how quickly it fell under gravity. They compared this to the force of the electric field required to lift it back up (against the force of gravity) to measure the magnitude of the droplet's charge. After many measurements it was found that every result was a multiple of the same number: 1.5×10^{-19} (about 1 percent out from the accepted value today). This showed that charge could not take just any value. Instead, it was the accumulation of huge numbers of electrons, and the oil-drop experiment had measured how much charge each one held. Like it or not, Millikan and Fletcher had showed that subatomic particles were here to stay.

Robert Millikan, shown here with the original oil-drop apparatus almost 20 years after the experiment, went on to be a founding member of Caltech, the California Institute of Technology.

64 Cloud Chambers

A WALK IN THE FOGGY SCOTTISH MOUNTAINS IN 1911 PROVIDED THE INSPIRATION FOR A DEVICE THAT TRACKED THE MOTION OF TINY PARTICLES. The curling streaks seen within this new "cloud chamber" would part the mists of the subatomic world.

The mountain specter is an optical illusion. The observer's shadow is cast on unseen diffuse clouds that are just in front of them. This would normally be more or less invisible if it were not for a bank of bright clouds far below the observer. The shadow shows up against this white cloud, making it appear as a giant figure in the distance. Fee fi fo ...

While hiking on Ben Nevis, the highest peak in Scotland, in 1911, physicist Charles Wilson was one of the lucky few to see the mountain specter. This is an optical phenomenon where the viewer's shadow appears as an immense figure looming on low clouds beside the mountain, haloed by rainbows. This got Wilson thinking about how clouds formed from droplets of water condensing around specks of dust or other nuclei.

Around ions

Back in his lab, Wilson had a go at recreating clouds by machine. He filled a flask with water vapor and controlled the temperature and pressure so they formed clouds of droplets, just like the real thing. He found that ions (charged atoms that have lost or gained electrons) made good nuclei for the droplets. Moreover, when a subatomic particle, such as an alpha particle, passed through the vapor, it bashed the electrons off the atoms in its path, leaving a trail of ions—and streaks of cloud. Wilson's cloud chamber had become a particle detector.

Other physicists added new features. The water vapor was replaced by alcohol fumes or carbon dioxide, and a magnetic field was added to deflect charged particles: The direction depends on the charge, while the angle of a deflection can be used to calculate mass. For the next 40 years, cloud chambers would give the clearest view into an increasingly crowded subatomic world.

A high-speed camera records the streams in a cloud chamber which reveal the trajectories of alpha particles.

65 Superconductors

It had been known for some time that the resistance of metals is reduced as the temperature of the conductor goes down. As researchers got better at chilling materials to ever lower temperatures, the question was whether resistance could be reduced to zero.

Electrical resistance is the product of matter getting in the way of a current of electrons. If that matter is jiggling around at high temperatures, it is more likely to get in the way of the electrons, and so the resistance is high. Removing heat energy from the conductor results in its particles moving more slowly and thus offer less resistance.

By the turn of the 20th century, scientists were getting very good at producing low temperatures. The method they used is broadly the same as the one still used in domestic refrigerators today. They make use of the Joule–Thomson effect, where a liquid refrigerant is allowed to expand very rapidly into a gas. The energy in the refrigerant is used to make its particles spread out, and so the individual kinetic, or heat, energy of each one is reduced. Heat from another source flows into the gas, making that source (the refrigerator compartment) become colder.

Liquid helium, solid mercury

In 1908, Dutch researcher Heike Kamerlingh Onnes succeeded in liquefying helium. No mean feat—its boiling point is 4.2 K (–268.95°C; –452.11°F). He then began using liquid helium as a refrigerant and was able to cool other materials to just above absolute zero. (By this time Walther Nernst had discovered the third law of thermodynamics: It is impossible to reach absolute zero.) In 1911, Onnes cooled mercury, a solid metal at this temperature, to 4.19 K. He found its electrical resistance disappeared entirely. The mercury had become a superconductor. The electrons in the metal were able to move completely freely without spending energy. This is a quantum effect seen on a large scale — electrons in orbit around a nucleus also experience zero resistance.

Superconductors have been developed that work at "high temperature"—around 130 K, although that is still very cold (–143°C; –226°F). Superconductors expel all magnetic fields, making it possible to make them levitate. This phenomenon is used in superfast "maglev" trains, which float above the track.

66 Cosmic Rays

WHILE CHARLES WILSON WAS INVESTIGATING IONIZED PARTICLES IN THE CONFINES OF HIS CLOUD CHAMBER, another physicist was looking for them among the real clouds. Up there he made a cosmic discovery.

An electroscope is a simple charge detector. It contains flimsy flaps of gold leaf which move in response to charge, repelling each other's similar charge. However, a fully charged electroscope eventually loses its charge. The reason could only be that charge—or electrons—is being lost to the air. And that means that the air has an opposite electric charge (albeit a slight one) to the foil.

The source of this charge was assumed to be the action of high-energy particles—naturally occurring from radioactive rocks and the like—which were ionizing the gases in the air. If that was the case then the charge should reduce with altitude, as the gases were getting farther away from the sources of ionizing radiation.

Victor Hess is photographed in his balloon prior to lift-off in 1911. He reached 5,300 m (17,388 ft) and returned with a discovery that would win him the Nobel prize in 1936.

Ups and downs

In 1910, Theodor Wulf, a German physicist and recent inventor of a highly sensitive electroscope design, carried one to the top of the Eiffel Tower, then the world's tallest building. He found that the charge of air was higher at the top than at the bottom, the opposite of the expected result. In 1911, Austrian Victor Hess set off on a series of high-altitude balloon flights to investigate further—and higher. Before takeoff he charged up one of Wulf's detectors, so its two gold foils were repelling each other. As the electroscope loses charge to the air, the foils move closer together as the repulsive force decreases. Hess confirmed Wulf's findings: The electroscopes lost their charge more quickly at higher altitudes. Therefore, the thin air up there was more heavily ionized than near the ground. Hess suggested that instead of coming from the surface of Earth, high-energy ionizing particles and radiation were coming from space! Hess's discovery eventually became known as "cosmic rays." Later investigations into cosmic rays would reveal them to be full of strange, exotic—and literally alien—subatomic particles.

67 The Quantum Atom

IN 1913, FOUR YEARS AFTER THE ATOM WAS FOUND TO BE A NUCLEUS SURROUNDED BY ELECTRONS, two young scientists revealed new ways of understanding the atom. Both based their findings on the way atoms gave out radiation. One became world famous, the other got shot in the head.

The two men, Dane Niels Bohr and Englishman Henry Moseley, made their mark while still in their twenties. Both spent time working with Ernest Rutherford, the godfather of nuclear physics, in Manchester. However, each had very different research interests.

Atomic number

Moseley was studying the X rays emitted by atoms. He found that atoms gave out a specific wavelength of X rays just as they produced certain colors of visible light. However, he also found that the X-ray wavelength was proportional to the amount of charge held by an atom's nucleus. Beginning with hydrogen as one, Moseley allotted an "atomic number" to the atoms of other elements according to the nuclear charges revealed by their X-ray spectra. This system tracked the atomic mass (or weight) of elements, as they were largely arranged in the periodic table: Helium had a charge of two on its nucleus, lithium three, and so on. Moseley's system proved to be a better way of organizing the periodic table of elements, which had originally been arrayed according to atomic weights and chemical properties. However, how it was that one atomic nucleus could hold more charge than another was still a mystery. Moseley never lived to find out. He was shot by a sniper during a World War I battle in 1915.

Niels Bohr, pictured in his Copenhagen laboratory ten years after his seminal work on the atomic model. In 1997, element 107, an enormously radioactive metal, was named bohrium in his honor.

Bohr's model

Moseley's work did not contradict the atomic model proposed by Ernest Rutherford. This stated that electrons orbited a nucleus and held in place by electromagnetic attraction. The combined charge of the electrons was equal and opposite to the charge of the nucleus.

Niels Bohr was interested in the motion and position of the electrons. At first he treated them like solids that moved according to the same laws of motion as a planet or pinball. He suggested that the kinetic energy

Like Planck before him Bohr's quantization of atomic physics came about through trying to create mathematical tools that could describe the observed behavior of atoms. The end result was an atom constructed of cloudy probabilistic waveforms rather than particles pinging around.

of an electron was proportional to the frequency of its orbit around the electron—in other words, how quickly it went around the atom. Bohr suggested that the constant of proportionality was a fraction of Planck's constant that linked energy and frequency of radiation.

When Bohr applied this view of the atom to the phenomenon of spectral emissions—the unique bands of light that each element produces—he found that he could only make the model work if electrons occupied certain positions, or orbitals, around the nucleus. It was not possible for them to be halfway between two positions.

Radiation of a specific wavelength (or frequency) contains a particular amount of energy. An electron moved from a low orbital to a higher one only once it was given the precise amount of energy. (There was no possibility of overshooting and falling back.) So an electron had to absorb a specific wavelength of radiation to receive the specific quantum of energy to make the jump—or quantum leap. Emission spectra resulted from the opposite process: Moving from a high orbital to a lower one gave out a specific wavelength of radiation.

ORBITAL SHAPES

Bohr's model of the atom did away with Rutherford's electron orbits, where electrons circled the nucleus like planets around a star. It replaced them with orbitals, energy levels that could hold specific numbers of electrons. The lowest level produces a spherical orbital, capable of holding two electrons, while higher-energy orbitals (see below) have the shapes of dumbbells and donuts and room for many electrons.

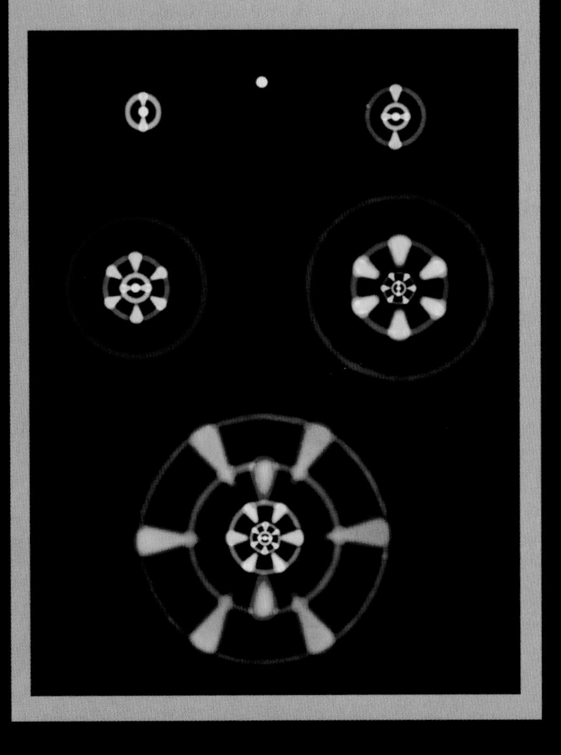

Complete with Scandinavian paneling, Niels Bohr built the first particle accelerator in Europe in 1939. He used it to bombard atoms with neutrons.

68 General Relativity: Space and Time

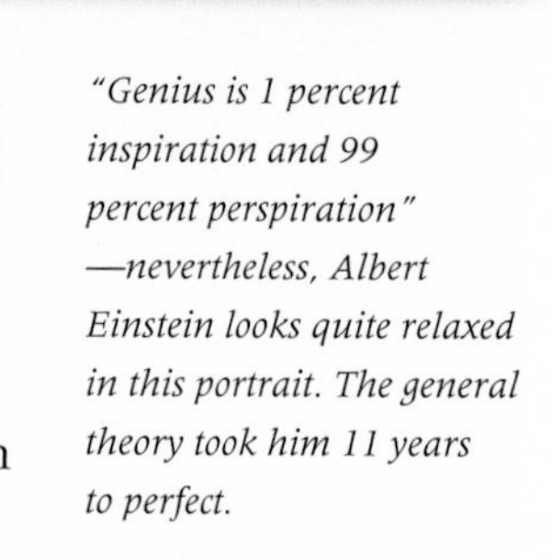

BY 1916, ALBERT EINSTEIN HAD EXTENDED HIS "SPECIAL" THEORY OF RELATIVITY TO A GENERAL FORM. This was able to correct the errors produced by classical mechanics when dealing with the motion of very large objects, like planets and stars.

Newton's laws of motion and gravitation work very well for predicting how a rock rolls down a hill, how a cannonball sails toward a target, and even how to fire a rocket at the Moon. However, when you use them to track the motion of huge objects like planets, small errors appear. In 1916, Albert Einstein's general theory of relativity provided the corrections. It demoted Newtonian physics from being an accurate model of how the Universe worked to a practical tool, a good shorthand for some simple problems (like flying to the Moon). If you really wanted to know what was going on, you needed general relativity.

Einstein's theory extended the link between space, time, mass, and energy that his special theory had made 11 years before. This time, he was able to show all of its effects, not just when traveling at or near to the speed of light. The shortest route from one point to another is always a straight line. However, straight lines can sometimes

"Genius is 1 percent inspiration and 99 percent perspiration" —nevertheless, Albert Einstein looks quite relaxed in this portrait. The general theory took him 11 years to perfect.

A picture taken by NASA's Hubble Space Telescope shows an "Einstein ring." The red galaxy at the center has warped space-time so much that light coming from a more distant galaxy located behind it has been curved into a horseshoe of light.

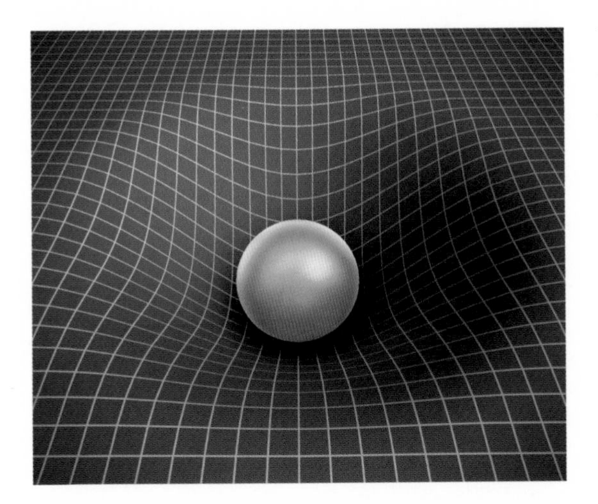

Space warps are not the stuff of science fiction. Every mass bends space-time.

be curved. This is because mass bends space-time, and heavy things like stars bend it a lot. If you used a tape measure (a very long one) to measure the distance between two points close to a star, the tape would appear to be just as straight as the line. However, the whole of space-time has been warped—including your tape. When viewed from outside the local area, even a beam of light shining between the points would be seen to curve.

Einstein was able to explain the force of gravity in terms of warping space-time. All mass bends space-time into a "gravity well"—the bigger the mass, the bigger this well. So a falling apple is plummeting into Earth's huge gravity well. Earth itself is inside the gravity well of the Sun, which is bigger still. However, instead of rolling to our doom, the planet's sideways speed is large enough to ensure we roll around the gravity well, like a roulette ball that never stops (not yet anyway). It took many years for relativity to be accepted by the mainstream, but today the whole of modern physics is built on general relativity and quantum theory.

69 The Proton

Henry Moseley was only 25 when he discovered why all nuclei have a specific atomic number. In 1917, this was found to be the number of protons in the nucleus. However, Moseley had died two year before, a casualty of World War I.

HENRY MOSELEY'S GROUNDBREAKING WORK HAD INDICATED THAT THE POSITIVE CHARGE OF AN ATOMIC NUCLEUS CAME IN COUNTABLE UNITS. Hydrogen nuclei had one unit of charge, so this was a good place to start the search for positively charged subatomic particles.

In 1815, William Prout suggested that the great variety of elements being revealed at the time had atoms made from multiples of hydrogen atoms. (Hydrogen is the lightest of all elements.) In 1917, he was proved to be partially correct.

This was the year that Ernest Rutherford discovered that blasting alpha particles at a nitrogen atom (atomic number 7), turned it into oxygen atoms (atomic number 8), and released a hydrogen nucleus. Rutherford knew that an alpha particle was a helium nucleus (atomic number 2). One of its units of positive charge had passed to the nitrogen nucleus, raising its charge to 8—and converting it to oxygen. That left one unit of charge, which turned out to be the same as a hydrogen nucleus (atomic number 1). Rutherford reasoned that positive charge was carried by a particle, which he named the proton, meaning the "first." Hydrogen had a single proton as its nucleus. The next discovery was that a proton had an equal and opposite charge to an electron but was almost 2,000 times heavier.

70 Wave-Particle Duality

LIGHT HAD BEEN SHOWN TO BE A WAVE, BUT IT WAS ALSO COMPRISED OF PARTICLES. MATTER, IN THE FORM OF ATOMS, HAD BEEN RECENTLY reduced to a set of smaller particles. Could it be that these were waves as well?

This is just what Frenchman Louis de Broglie was thinking in 1923. That year he proposed that wave-particle duality was not only a feature of light and other forms of energy. The same phenomenon could be applied to matter—particles, like electrons and protons, which have mass. He came up with the concept of a matter wave. Every particle of matter can be treated as a "waveform." This does not propagate like a light wave, but shares some of the same features: The speed of the particle is inversely proportional to its wavelength—faster particles have shorter wavelengths. The kinetic energy in the particle is proportional to its frequency.

Particles are waves

De Broglie used math to back up his proposal, but in 1927, George Thomson, son of J.J, the man who had discovered the electron 28 years before, found some tangible evidence for it. Thomson repeated the Young experiment, which had shown light to be a wave, but used a beam of electrons instead. He fired the electrons at a screen with two slits in it. A detector behind the screen then picked up each electron that passed through, representing it as a black dot. If electrons were simply particles they would form two groups of dots, one behind each slit. However, Thomson found the dots formed into the same dark stripes of an interference pattern, like you would expect waves to produce. Thomson had found that his father's subatomic particle was not only a particle—it was a wave and a particle at the same time.

STARK EFFECT

In 1913, Johannes Stark (below) found that electric fields caused a spectral line (an emission of a specific frequency of radiation) to split into several lines. This was due to the field altering the waveform of the electron, and offered a way of probing the properties of electrons.

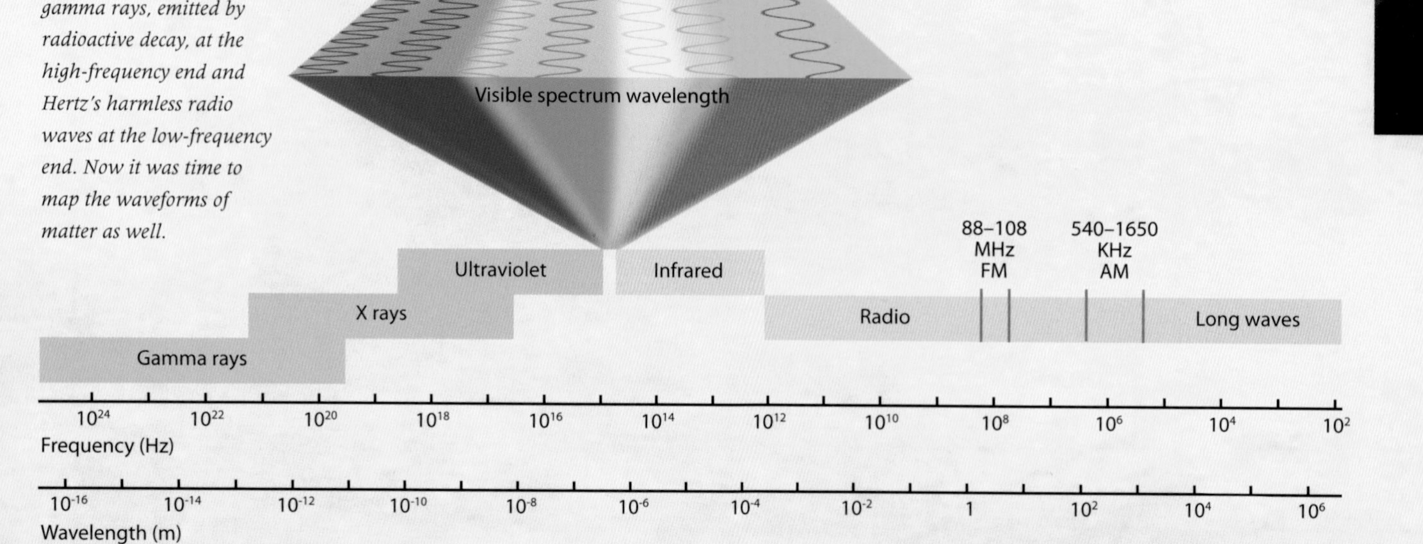

By the 1920s, the full electromagnetic spectrum had been mapped out with gamma rays, emitted by radioactive decay, at the high-frequency end and Hertz's harmless radio waves at the low-frequency end. Now it was time to map the waveforms of matter as well.

71 Exclusion Principle

ARMED WITH THE MATHEMATICAL TOOL OF WAVEFORMS, A NEW BREED OF QUANTUM PHYSICISTS BEGAN TO EXPLORE THE PROPERTIES OF PARTICLES. In 1925, Wolfgang Pauli found some of them were rather exclusive.

Austrian physicist Pauli was attempting to make head or tail of all the possible combinations of properties that the electrons in an atom could have. He realized that the data all pointed to each electron occupying one of a fixed number of possible energy states. This is now equated as "spin," a quantum property of a particle related to its angular momentum. Pauli said that no two electrons in the same atom have the same spin. This became the exclusion principle and it was later extended to all particles with a half-integer spin (a value that is a multiple of 0.5). At the quantum scale, no two of this set of particles can be in the same place at the same time.

72 Bosons: Force Particles

NOT EVERY QUANTUM PARTICLE HAD A HALF-INTEGER SPIN; SOME HAD WHOLE NUMBER SPINS. THESE PARTICLES WERE EXEMPT FROM THE EXCLUSION PRINCIPLE, and could occupy the same space and energy. Welcome to the bosons.

In 1925, Satyendra Nath Bose, a Indian physicist, collaborated with Albert Einstein to produce a set of rules that defined how particles with whole-integer spin behaved. Such particles include the photon, which is the force carrier for electromagnetism: Opposite charges attract each other (and like ones repel) because photons are ferrying energy between them. Force-carrying particles were named "bosons" in honor of Bose. The half-integer spin particles were named "fermions" for Enrico Fermi (more of him later.)

Quantum physics was steadily learning how to express a growing family of particles in terms of math, creating complex vibrating waveforms, where once were solid spheres.

73 An Uncertain Universe

THE NEW FIELD OF QUANTUM MECHANICS WAS SHINING A LIGHT ON THE NATURE OF MATTER AT ITS MOST BASIC LEVEL. However, by 1927 it appeared that there was a limit to what we could know about this realm of reality. The quantum universe was riven with uncertainty.

The bearer of this uncertain news was Werner Heisenberg, a young researcher at Niels Bohr's Institute of Theoretical Physics in Copenhagen. Soon after, the world's top physicists gathered at the Solvay Conference in Belgium to discuss the way forward for quantum theory. The debate focused on imposing limits on when researchers would be able to say that they had actually discovered something. Quantum uncertainty would have a big impact on this. Albert Einstein was not a fan of the concept, saying that "God does not play dice." Bohr responded: "Einstein, stop telling God what to do!"

Werner Heisenberg's name is now synonymous with one of the counterintuitive features of quantum physics: Uncertainty.

Uncertainty principle

In the mid-1920s, Heisenberg and a fellow quantum physicist, Max Born, had found that probability, the field of math that deals with chance, was central to the waveforms of quantum particles. This eventually

HALF-LIFE

Quantum uncertainty ensures it is impossible to predict when a radioactive atom will decay. Instead, the rate of decay is expressed in terms of probability. Highly unstable elements are likely to break down faster than less radioactive sources. The decay is represented as a half-life. This is the time in which half of the atoms are likely to decay. The half-life of the most common form of uranium is 4.46 billion years, about the age of our planet. That means that half of the uranium that was present as Earth formed has now gone.

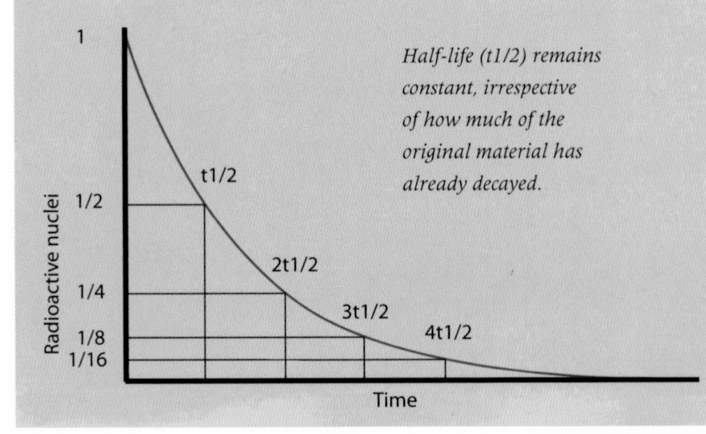

Half-life (t1/2) remains constant, irrespective of how much of the original material has already decayed.

Schrödinger's cat is the name for a thought experiment proposed by Erwin Schrödinger in 1935. Its purpose is to highlight the observation effect. It puts a cat, a hypothetical one, in a box. The box contains a poison that is released when triggered by the decay of a single radioactive atom—an entirely unpredictable and random event. Schrödinger said that with the lid closed it is impossible to know if the cat was alive or dead. However, in quantum theory it is possible for it to be equally alive and dead until an observer checks.

became the Uncertainty principle: The more precisely the position is determined, the less precisely the momentum is known in this instant, and vice versa. This is not a problem with measuring devices, but a property of wave functions. Pinning down the speed, or more accurately, the momentum (mass times velocity) of a quantum particle means that its location can only be expressed in terms of a probability—the chances of it being in one place or another.

Causality

This gives quantum particles multiple states—they can be in more than one place at a time—and only settle into a definite one when you look. This idea is called superposition. Such a thought disrupts the link between cause and effect. One of the central tenets of physics is that the present is the result of events in the past, and the events taking place now will cause the future. Superposition of the particles in a quantum system breaks this link. Effects at this level can occur without any cause, and there is no certainty what will happen. Before an observer measures it, every outcome is just one of many possible futures. As well as making everyone's head hurt, superposition could form the basis of superfast computing and teleportation in future. But that is far from certain.

74 Geiger's Counter

A CLICKING MACHINE THAT THEN GIVES AN ALARMING SQUEAK IS A FAMILIAR TROPE IN TV THRILLERS. This is the Geiger counter, a radioactivity detector developed by a man who was there at the birth of nuclear physics.

Hans Geiger, with guidance from Ernest Rutherford, put together the first radioactivity detector in 1908. Initially, it was only capable of picking up alpha particles but by 1928, Geiger, with help from his research assistant Walther Müller, had enhanced the device enough so it could pick up all kinds of high-energy radiation. Properly termed a Geiger–Müller tube, the detector is a sealed tube filled with a low-pressure gas. There are two charged electrodes inside, but the high resistance of the gas blocks a current from running between them. When high-energy radiation enters the tube, it ionizes the gas (charging its molecules), and that allows a pulse of electricity to flow. The pulse rate equates to a measurement of the amount of radiation in the area. A pulse of electricity through a speaker makes a click—and a lot of clicks together make a squeak.

This copper Geiger–Müller tube from 1932 was used in the discovery of the neutron.

75 Antimatter: The Same But Different

IN 1928, PAUL DIRAC PRODUCED AN EQUATION THAT ENCOMPASSED ALL THE QUANTUM PROPERTIES OF THE ELECTRON. Not only did this bit of math work perfectly in describing the behavior of the negatively charged electron, it also suggested a positively charged version could exist.

$$(c\alpha \cdot \mathrm{p} + \beta mc^2)\,\varphi = i\hbar \frac{\partial \varphi}{\partial t}$$

The Dirac equation expresses the waveform of a half-spin particle. At the time the only particle known to have this feature was the electron, but now there are 11 more.

In 1928, the subatomic family consisted of negatively charged electrons, positively charged protons, and uncharged, or neutral, photons—the boson particles that mediated the force of attraction between the other two. This was what made up matter. The Dirac equation suggested that a positively charged electron would work just as well—but require a negatively charged proton. This mirror-image matter, identical in every way but its charge, was dubbed "antimatter." But did it exist? If it were to, then what would happen to it? The theory suggested that matter and antimatter annihilate each other, resulting in nothing but a release of energy. If antiparticles were out there, then they did not last for long.

As had been the case with other breakthroughs in quantum theory, Paul Dirac's mathematics revealed a lot more about the Universe than it had been intended for.

76 Atom Smasher

ERNEST LAWRENCE WANTED TO DO BIG PHYSICS. WHILE HIS CONTEMPORARIES OFTEN WORKED WITH THEORY, LAWRENCE WANTED TO SEE ATOMIC ENERGY in action. And so he invented the world's most powerful particle accelerator.

As the atomic age set in, Ernest Lawrence was an unusual figure among the world's top physicists: He was an American, in a field dominated by European-born scientists. Lawrence had a typically New-World attitude to science—he wanted to do it on a very big scale. One way to find out about the inner workings of the atom was to smash them together and see what came out. In 1929, Lawrence invented the cyclotron, a particle accelerator that spiraled particles up to high speeds before hitting a target.

Other types of particle accelerators had been built already. Those were linear (straight-line) tracks that fired a beam of ions through a series of oscillating electric fields. Each field accelerated the ions a little. (Ions are atoms that have lost or gained electrons, so they carry a charge making them susceptible to electromagnetic forces.)

In a whirl

Lawrence's cyclotron, or "atom smasher," also uses electric fields to accelerate ions. However, his device keeps the beam going around in circles, so the push from the electric field is more or less constant, and that results in higher speeds than were previously possible. A cyclotron has two D-shaped electrodes that make a circle with a small gap across it. An electric field runs across the gap between the dees. This entire structure is surrounded by a powerful electromagnet. Its magnetic field is perpendicular to the electric field, running down through the dees. A sample of ionized material is injected into the center of the machine and it is pushed into one of the dees by the electric field. However, the magnetic field makes the material swing around so it continues to speed up, this time heading over to the other dee. As the ions are accelerated to greater speeds, the circular path they take grow wider, and they trace a spiral moving out to the edge of the dees. They are now at top speed and therefore directed out of the cyclotron into a side chamber to collide with the target.

Lawrence's first model was 10 cm (4 in) wide but achieved a speed of 1 percent of the speed of light. Wider cyclotrons could achieve faster speeds. Lawrence's final cyclotron was 467 cm wide (184 in) and was more than 1,000 times as powerful as the first.

TECHNETIUM

In 1935, Element 43 was missing. It is so radioactive and unstable that hardly any is left on Earth. In 1936, two physicists from Italy asked Lawrence to examine the used parts of an old cyclotron. On them, they found atoms of element 43! They had been made by ions of other elements colliding in the cyclotron. Because it was the first element to be made in a lab it was named technetium.

43
Tc
TECHNETIUM
(98)

Ernest Lawrence (bottom) readies a cyclotron in 1938 with the help of Donald Cooksey. The accelerating dees are located behind Lawrence. He and Cooksey are working on the target chamber that receives the beam of high-speed ions.

77 The Electron Microscope

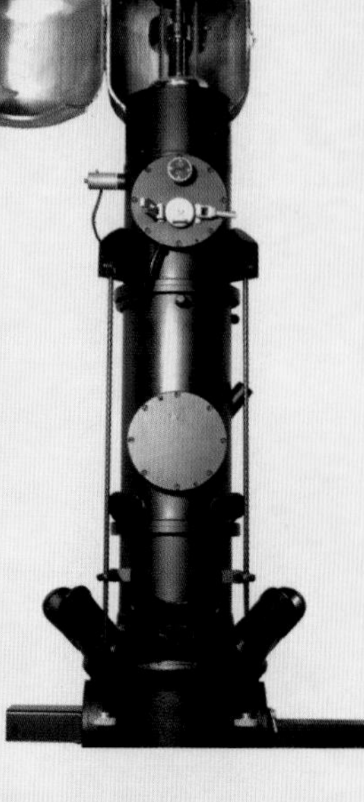

A vintage electron microscope showing the electron gun positioned at the top.

WHEN RENAISSANCE-ERA PHYSICISTS LEARNED TO FOCUS LIGHT WITH LENSES INSIDE TELESCOPES AND MICROSCOPES, THEY EXTENDED THE REACH OF THE human senses. In the 1930s, they did the same with electrons.

Optical equipment makes use of the wave nature of light. Now that electrons had been shown to behave as a wave as well, did that mean they could be focused into images? Light microscopes have a limit; they cannot see objects below around 200 nanometers. Light's wavelength is too large to resolve images smaller than that. Electrons have a wavelength 100,000 times shorter than visible light. That means an electron microscope could see objects that are 50 picometers (trillionths of a meter)! Electromagnetic lenses had been developed in the 1920s, and the following decade the first electron microscopes (EMs) were being tested. A transmission EM forms an image from how a beam of electrons is scattered as it passes through a sample. Scanning EMs do it by reflecting electrons off specimens.

78 Neutrons: The Final Piece

James Chadwick's neutron-weighing apparatus.

SOMETHING WAS NOT ADDING UP IN THE ATOM. PROTONS AND ELECTRONS WERE NOT enough to account for all the matter inside.

Atoms had been shown to have a specific atomic number. This was shown to be the number of units of positive charge in the nucleus, and later found to be the number of protons in the atom. However, atoms also had another value, that of atomic mass. This was a relative measure, with hydrogen, the lightest atom, having an atomic mass of 1. The mass of all other elements was expressed as a multiple of hydrogens.

Hydrogen has a nucleus of just a single proton. The atomic number and atomic mass are equal. However, that is not the case with any other element. By the 1920s, it was a widely held belief among atomic physicists that the extra mass of the nucleus was due to the presence of an uncharged particle. Although atomic mass was a much less consistent measure—the atoms of the same element come in a range of masses called isotopes—it was generally about double that of the mass of protons alone. Perhaps the unknown neutral particle was the same mass as a proton?

In the early 1930s, researchers found a new type of radiation that was produced when high-speed alpha particles hit samples of beryllium or boron. This radiation had no charge but was far more powerful than gamma rays. In 1932, James Chadwick directed the radiation into a variety of gases made of molecules of different masses. He measured the displacement of the gases to calculate the mass of the particles in the radiation, and found it was about the same as that of protons. Here was the neutron—a neutral particle that completed the atomic nucleus.

79 Positrons: A New Puzzle

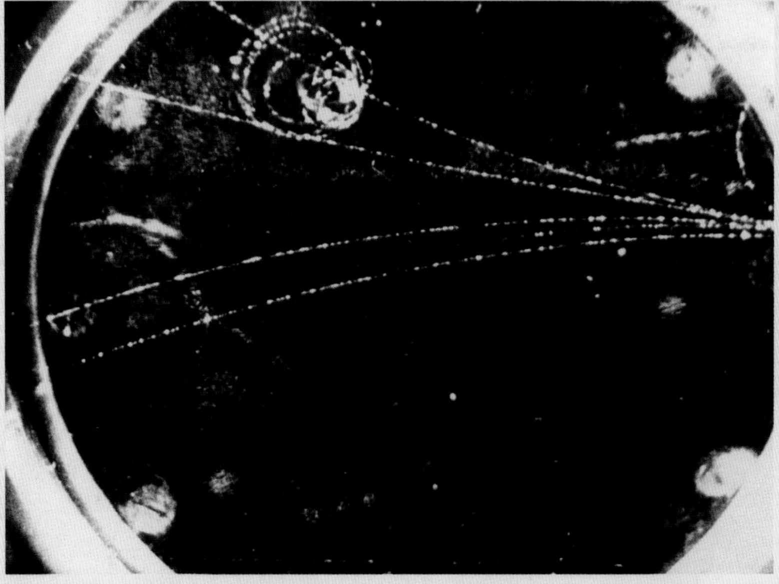

A cloud trail shows electrons and positrons parting ways as they are deflected in opposite directions by the cloud chamber's magnetic field.

WITHIN MONTHS OF NEUTRONS SOLVING THE MYSTERY OF ATOMIC MASS, the first evidence emerged of antimatter. A new puzzle had been found.

In 1929, researchers capturing particles produced by high-energy cosmic rays hitting the atmosphere noticed what looked like electrons, except they had a positive charge. The ramifications of Paul Dirac's proposal of antimatter the year before were still being digested by the scientific community, so these anomalies were disregarded. However, when Carl Anderson saw the same thing in his cloud chamber in 1932, he knew what he was looking at. He named the particle a positron, a portmanteau of "positive electron."

These positrons were just fleeting visitors to our planet. They were formed when larger particles (probably atomic nuclei) arriving from space at nearly the speed of light, hit the gases in Earth's atmosphere. Such collisions are powerful enough to create particles from pure energy. They then release positrons as well as a whole family of "exotic" particles that were yet to be discovered.

80 Missing Matter

DURING THE 1920S, THE DIMENSIONS OF THE UNIVERSE HAD BEEN ROLLED BACK TO A SIZE NEVER IMAGINED BEFORE. And Edwin Hubble had shown that it was getting bigger. Then, in the 1930s, a shocking discovery was reported : Most of the Universe is missing!

WIMP v. MACHO

We are still in the dark on dark matter. There are two broad descriptions of what it could be: WIMPs (Weakly Interacting Massive Particles), which have mass but do not interact with detectors, and MACHOs (Massive Astrophysical Compact Halo Objects), which are big dark things like black holes and cold stars, anything too dark to see.

Fritz Zwicky got his first evidence for dark matter from the rotation of some of the hundreds of galaxies in the Coma Cluster.

Astronomers use spectroscopes to see what stars are made of. The gases around a star absorb certain colors of light, so starlight is filled with gaps. Each gap corresponds to an element. However, the normal wavelengths had shifted out of position. This was the Doppler effect writ large. Shifts toward blue meant the galaxy was moving toward us. Redshifts were much more common, and indicated objects moving away. In 1929, Edwin Hubble helped to prove that everything beyond our immediate intergalactic neighborhood is moving away from everything else: The Universe is expanding.

In a spin

Astronomers were interested to know what part gravity was playing in this expansion. The force of gravity is an attraction between all matter, so how come matter was moving apart? The big question was how much matter does the Universe contain. In 1932, Jan Oort found that the Milky Way was spinning too fast for the amount of material in it. The following year, Fritz Zwicky saw the same thing in the motion of other galaxies. He reasoned there was more material in them than he could see, and named the invisible stuff *dunkle materie*, better known today as "dark matter." The theory was—and still is—that the dark patches of the Universe are not empty "space." Instead, it is filled with material that does not give out light. The only thing dark matter really does is contribute to the gravitational field. In the 1970s, dark matter was weighed by observing how huge masses of visible matter bent space and light. It was found that there is five times as much dark matter than visible matter!

Fritz Zwicky ran the Palomar Observatory in California and searched for supernovae—giant exploding stars.

81 Indoor Lightning

BY THE 1930S, ELECTROSTATIC GENERATORS HAD MOVED ON SINCE THE DAYS OF VON GUERICKE'S SPINNING BALL of sulfur. Robert Van de Graaff's powerful design was enough to make your hair stand on end.

Seeing a Van de Graaff generator in action is generally one of the more memorable events in physics class. Everyone will have seen a long-haired person touching a metal sphere with their hair sticking out in all directions. The sphere's charge is transferred to the person, and every strand of hair repels the others, creating that memorable hairstyle.

The American inventor Van de Graaff, was not intending to build a novelty. Back in 1929, he conceived of the generator to make huge voltages for use in linear particle accelerators. The immense electric fields produced by his huge machines were used to ionize particles and accelerate them at great speed into detectors.

The Van de Graaff generator works in a similar way to the friction-powered devices from the early days of electrical research (like the sulfur ball). A moving belt inside the generator's column picks up charge by rubbing against an electrified comb contact. The charge is carried up to a dome, where the charge jumps from the belt to a similar contact. As a result, the dome can collect huge voltages and give out bolts of artificial lightning.

MIT's Van de Graaff generators were built in a disused airship hangar in 1933. They could generate 10 megavolts, which was used to zap particles positioned in a tube between the domes. Incredibly, both domes contained laboratories which were occupied during live experiments! Being inside the dome was safe, although the same cannot be said for birds flapping between the charged towers.

82 Speeding Light: Cherenkov Radiation

When a Russian researcher saw an eerie, blue light glowing from a bottle of water under radioactive bombardment, he realized that there was a way for things to travel faster than light.

Let's be straight, light travels fastest in a vacuum, and that is the upper limit of speed—nothing beats it. However, since the days of Snell's law, it had been known that light slows down as it moves into different transparent media, like air or water.

In 1934, Pavel Cherenkov figured out what was making the water glow. The radioactive particles that were being blasted into the water were breaking the speed limit for that medium. The speed of light in water is just three-quarters of what it is in a vacuum. However, it is possible for particles of matter to travel through water faster than light can (although in a vacuum, matter never matches light for speed).

Cherenkov radiation, coming from nuclear fuel rods, makes the coolant water in a reactor glow blue.

Cherenkov effect

As these speeding particles—mostly small things like electrons—tear through the medium, they disrupt the surrounding atoms, making them radiate energy. However, the slowed light cannot shine away from the super-fast particles. When air cannot move away from a supersonic jet breaking the sound barrier, it forms a sonic shock wave. In the Cherenkov effect, something similar happens: A kind of shock wave of photons develops, which creates the blue glow.

Detection system

The Cherenkov effect provides valuable evidence of fast-moving particles arriving from space. (These are termed cosmic rays despite being particles moving at incredible speeds.) A flash of Cherenkov radiation heralds their arrival, as they collide with air, creating jets of high-energy particles. Cherenkov detectors are also on the lookout for the elusive neutrinos that flood through our planet.

83 Exotic Particles

THE COLLISION OF COSMIC RAYS WITH THE ATMOSPHERE WAS FASTER THAN ANY PARTICLE ACCELERATOR AROUND IN THE 1930S. Although they were difficult to observe, these high-energy collisions were the best place to find a new set of short-lived subatomic particles.

Hideki Yukawa was awarded the 1949 Nobel prize for physics for his work on mesons.

Japanese theoretical physicist, Hideki Yukawa, was interested in what kept the atomic nucleus together. The force of electromagnetism would push the positively charged protons apart (like charges repel each other), so there must be a stronger force that overrides electromagnetism and holds protons and neutrons together. (Known as the strong interaction, this force only works across distances much smaller than a width of an atom—it does not reach out to where the electrons are circling.)

In 1934, Yukawa suggested that the strong force was mediated by medium-sized particles that were halfway between an electron and proton (or neutron). He named them mesons. In 1936, the mu meson was discovered in cosmic-ray particle jets. However, this did not match Yukawa's theory and was later renamed as a muon—it is a kind of oversized electron. In 1945, a pi meson was seen in cosmic-ray collisions. It had two-thirds of the mass of a proton and was the first true meson. This "pion" acts as a "virtual" particle: It exists for a very small length of time (measured in billionths of a second) before decaying. In the end pions and a host of other mesons would prove to be a stepping stone to even more fundamental particles called quarks and gluons.

The spiral tracks in a bubble chamber reveal kaons (K mesons), which are involved in both the strong interaction and weak force of radioactive decay.

84 Superfluidity

WHEN A LIQUID GETS VERY COLD, JUST A FEW DEGREES ABOVE ABSOLUTE ZERO, SOMETHING VERY ODD HAPPENS. It stops being limited by its container, creeping up and over the sides, and eventually emptying all by itself. It has become a superfluid.

Superfluidity was first found in a sample of liquefied helium in 1937 by teams working in Russia and Canada. Helium becomes a superfluid at just above 2 K (that's –271°C or –456°F). To look at, the liquid looks like any other, but it has zero viscosity. That means it has no inherent stickiness that stops it changing shape. A superfluid can also flow without friction. At this temperature, the liquid is displaying quantum effects but at the macroscale (as is the case in superconductors). Helium is especially good at this because most helium nuclei have four particles. That allows them to behave like bosons at low temperatures, becoming exempt from the exclusion principle.

Superfluid helium escapes from a cup by creeping up and over the sides. Most liquids cling to the side of their containers but their viscosity prevents them from climbing too high—not so with superfluids.

85 Nuclear Fission

Chicago Pile 1, the first nuclear reactor, was built on some disused racketball courts behind the west stand of the University of Chicago's sports stadium.

IN THE 1930S, PHYSICS MET POLITICS IN A BIG WAY, AS IT WAS REVEALED THAT SCIENCE HAD FOUND A MEANS OF TURNING matter into energy—by splitting the atom. It all began with the discovery of the neutron ...

Enrico Fermi, an Italian physicist, wanted to use the neutron as a means of probing the atom, just as Rutherford had used alpha particles to probe the atomic nucleus. Unlike alpha particles, neutrons have no charge so they would not be deflected by the charges present inside atoms. Fermi's team in Rome set about bombarding all kinds of atoms with their new tool. In 1934, they announced that an element with an atomic number of 94 (Fermi named it hesperium) had been formed from uranium. No one was quite sure how this was possible. (It's now called plutonium.) The German Otto Hahn repeated the neutron bombardments experiments and found barium in a sample of uranium. His colleague Lise Meitner showed that the barium had formed because a neutron had entered an uranium nucleus, making it very unstable. Instead of decaying in a normal way and emitting small particles, the nucleus had just split in two. This was nuclear fission.

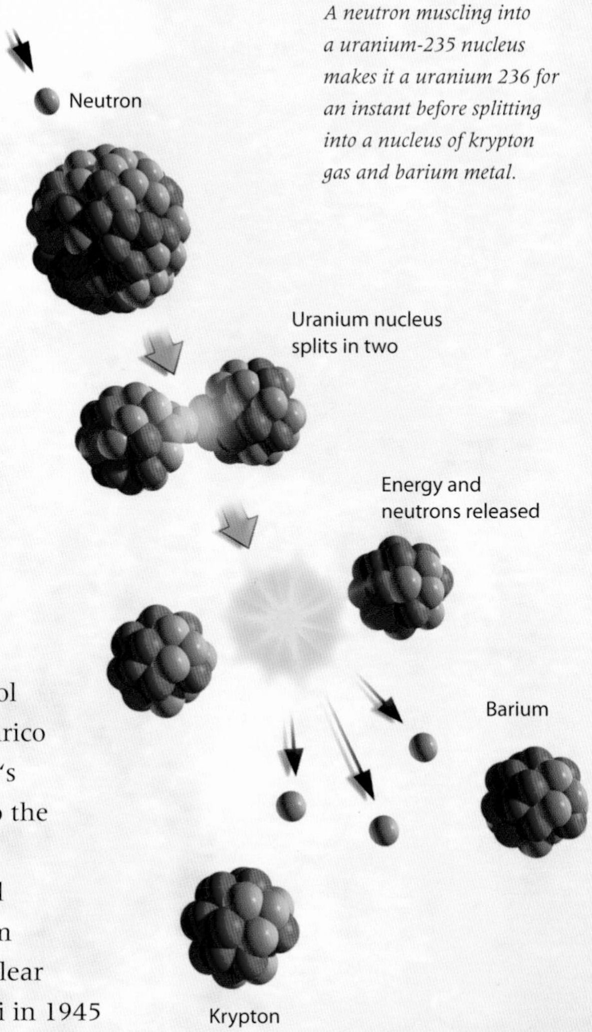

Chain reaction race

As Einstein's energy-mass equation ($E=mc^2$) predicted, fission reactions released huge energy. Leó Szilárd, a Hungarian working in America, found that if a fission released two or more neutrons as the nucleus split, then a chain reaction could result. Each splitting atom resulted in at least two more fissions until all the atoms had been used up. Uncontrolled, this could be used in a terrible bomb. With Europe on the brink of war, Szilárd and Fermi (now in New York to escape the Nazis) decided to keep quiet about this development. However, Frédéric Joliot-Curie (Marie's son-in-law) also figured it out. In 1939, he reported that a nucleus of uranium-235 split to produce at least three neutrons.

As the world went to war, the race was on to learn to control fission, and so perhaps create the decisive weapon. In 1942, Enrico Fermi set up Pile 1, the world's first nuclear reactor, at Chicago's university. Blocks of graphite were used to focus neutrons onto the uranium and so create a slow chain reaction.

The Manhattan Project followed Fermi's research. It focused on "enriching" uranium, boosting the proportion of U-235 from 0.7 percent in natural samples to a high enough level for a nuclear explosion. The atomic bomb strikes on Hiroshima and Nagasaki in 1945 proved to the world that nuclear fission was a force to be reckoned with.

Neutron

Uranium nucleus splits in two

Energy and neutrons released

Barium

Krypton

A neutron muscling into a uranium-235 nucleus makes it a uranium 236 for an instant before splitting into a nucleus of krypton gas and barium metal.

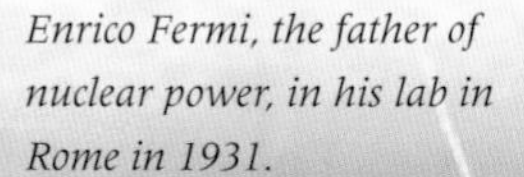

Enrico Fermi, the father of nuclear power, in his lab in Rome in 1931.

MODERN PHYSICS

86 QED: Quantum Electrodynamics

FEYNMAN DIAGRAMS

Simple squiggles and arrows are the starting point for QED. The straight lines show the path of electrons, and the wavy ones are photons being absorbed and emitted by electrons. The diagrams serve as visualization of an interaction that has a certain probability of occurring.

***QUOD ERAT DEMONSTRANDUM*, OR QED, IS A SCIENTIFIC WAY OF SIGNING OFF A DISCOVERY. IT IS SHORTHAND FOR "WHICH HAD TO BE DEMONSTRATED"** and shows something has been proved. In physics, the QED theory proves more than anything else.

To a physicist, the initials QED stand for quantum electrodynamics. One of the founding figures of this branch of physics, American Richard Feynman, has described it as the "jewel of physics." No other theory is able to demonstrate the same phenomena all the way from the quantum level of subatomic particles to the relativistic scale, that of stars, galaxies, and black holes.

Electrodynamics really begins with James Clerk Maxwell's field equations and was developed by Albert Einstein, who introduced the idea of the photon, a particle of light, and explained how light behaved as it traveled vast distances across space-time with his special theory of relativity. But then the initial messages coming from quantum theory was that these "classical" theories from the macroscale Universe could not be directly linked to what was happening inside atoms.

Richard Feynman was a master communicator as well as a Nobel-prize-standard physicist. He was able to convert complex quantum events into simple snapshots.

Many steps

The first glimmer that such a link could be made came in the 1920s with Dirac's electron equation. This showed how energy and matter interacted to produce light and other electromagnetic radiation.

However, the complexities of quantum particles were an obstacle to creating a general theory about how light behaved when it met fundamental particles of matter. Then in 1947, Hans Bethe

(soon to be linked to the Big Bang theory) found a mathematical technique to cut through the chaos of quantum variables. This proved to be a springboard for other physicists, Richard Feynman among them, to come up with a series of ways to explain electromagnetic phenomena on the quantum scale.

Many steps

At first several different approaches were used, but they were all later shown to be the same. Richard Feynman's diagram-based system has become an integral part of the theory, helping to demystify the most cutting-edge physics. They are so versatile, Feynman diagrams are now used to represent the interactions of all quantum particles.

Despite the fun visuals, at its heart QED contains some very tough mathematics, which allows physicists to give accurate probabilities of specific interactions between electrons and photons. And it works in predicting the behavior of different particles that before were an unfathomable foam of possibilities.

87 Transistors

THE EARLY 20TH CENTURY SAW THE EMERGENCE OF ELECTRONICS, a technology that uses a series of "switches" to carry out mathematical instructions. In 1947, a big breakthrough was made.

The first working transistor was a far cry from the tiny components driving modern computing technology.

A computer needs a program, even the early versions being developed in the 1930s and 40s. The program is inputted as a series of 0s and 1s, which switch components on and off in a specific way to process data and perform tasks.

In 1947, three researchers at the Bell Labs invented a new type of switch that would change the world—the transistor. It was developed to replace thermionic valves, tubes of gas which could be made to conduct or block electricity. Transistors did the same but made use of a semiconductor, a material like silicon that could act as both a conductor and an insulator. At the time, quantum physics was revealing how electrons behaved to make a material switch from one to the other. The semiconductor transistor, miniaturized onto a microchip, would go on to change the world.

88 The Big Bang

IF THE UNIVERSE IS EXPANDING, THERE MUST HAVE BEEN A TIME WHEN IT WAS MUCH SMALLER THAN IT IS NOW. TAKING THAT LOGIC TO ITS CONCLUSION, space must have started out occupying a single point. How did we get from that point to where we are now?

The theory that explains how we know that the Universe is expanding is called Hubble's law. In 1912, Vesto Slipher discovered that distant objects have larger redshifts. (A redshift is when the light's wavelengths have been increased by the Doppler effect, and the color has shifted toward red. It shows an object is moving away from us.) In 1929, Edwin Hubble showed that not only were galaxies moving away from us, but they were also moving away from each other. Space itself was expanding, stretching the light from more distant galaxies more than the ones nearer to us.

Hubble is quite rightly hailed as a hero for this great step forward in science, but a Belgian priest, Abbé Georges Lemaître, had come to the same conclusion two years earlier. He used Einstein's theory of relativity to show that the Universe could never be static—stay the same—and had to be dynamic, either contracting or expanding. In 1931, Lemaître suggested that the dynamic Universe had begun in an almighty explosion from a single point, which he called the "primeval atom." Lemaître's idea

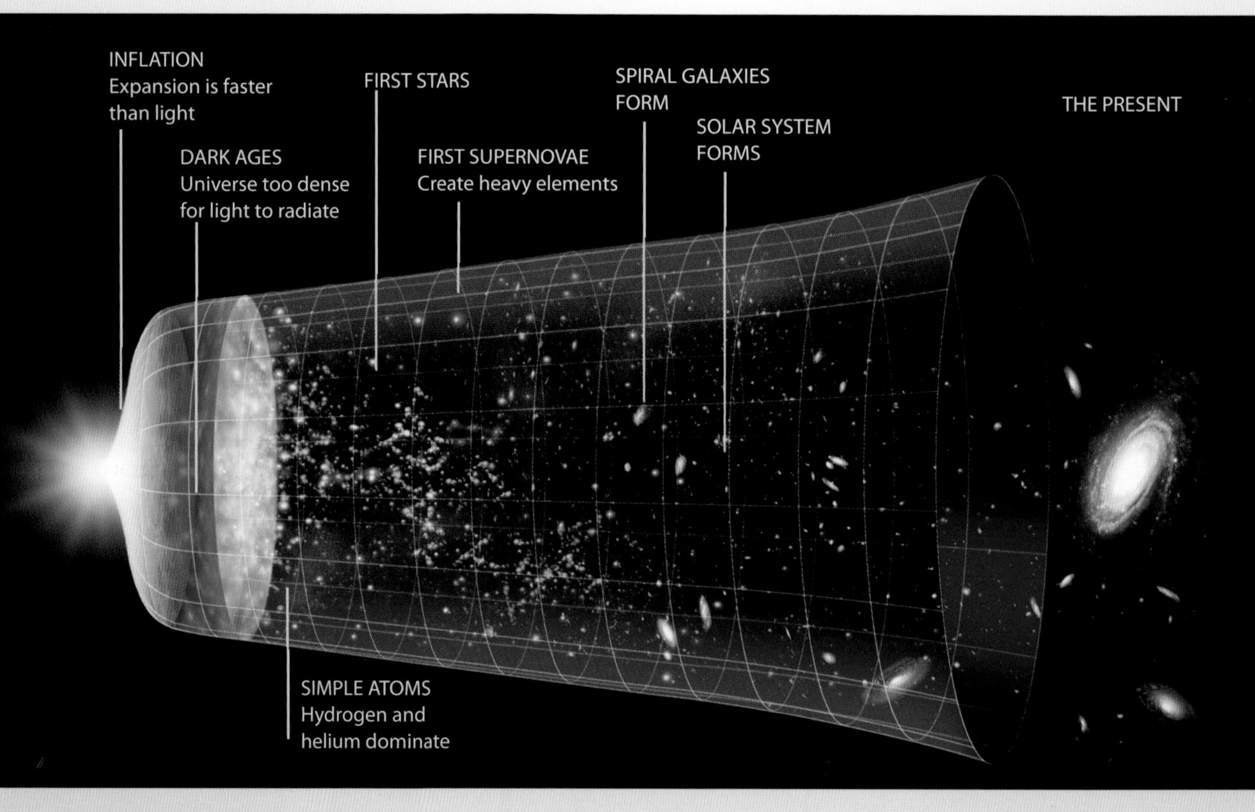

BANG!

The Big Bang was so big that it happened everywhere, all at once. The whole of space was incredibly hot. But back then space was very small. As it grew in size, it cooled down. This allowed energy to form matter, matter to form atoms, atoms to form stars, and eventually our planet. The farther we look into the Universe the older the light we are seeing. The farthest we have seen is 13.2 billion light-years. The Universe is 13.8 billion years old.

The Origin of Chemical Elements

R. A. Alpher*
Applied Physics Laboratory, The Johns Hopkins University, Silver Spring, Maryland

AND

H. Bethe
Cornell University, Ithaca, New York

AND

G. Gamow
The George Washington University, Washington, D. C.
February 18, 1948

As pointed out by one of us,[1] various nuclear species must have originated not as the result of an equilibrium corresponding to a certain temperature and density, but rather as a consequence of a continuous building-up process arrested by a rapid expansion and cooling of the primordial matter. According to this picture, we must imagine the early stage of matter as a highly compressed neutron gas (overheated neutral nuclear fluid) which started decaying into protons and electrons when the gas pressure fell down as the result of universal expansion. The radiative capture of the still remaining neutrons by the newly formed protons must have led first to the formation of deuterium nuclei, and the subsequent neutron captures resulted in the building up of heavier and heavier nuclei. It

The Alpher, Bethe, and Gamow letter put the Big Bang theory on a more scientific footing.

was very intuitive. He had no evidence to back it up, but suggested that the primeval atom shattered into all the atoms in the Universe—spreading out in all directions.

That "big bang"

The theory sounded good but lacked detail. In 1948, George Gamow (Gammov) and Ralph Alpher, two colleagues fresh from the Manhattan Project, proposed a more rigorous method of nucleosynthesis (nucleus formation). (Gamow decided to add his friend Hans Bethe to the list of authors to make Alpher, Bethe, and Gamow—a pun on the first three Greek letters.) Their suggestion was that visible matter of the Universe had developed through the continual fusion of particles into more complex and massive forms. This process occurred as the Universe cooled and spread out. Ironically, one of the theory's chief opponents gave it its name. In 1949, Fred Hoyle, an eminent astronomer, described it as a "big bang,"contrasting it with his alternative, the now-invalidated steady-state theory which proposed that matter is being added to the Universe continuously as it expands.

89 Bubbles and Sparks

By the 1950s, two new types of particle detector were in use. Their extra sensitivity was much valued as they were able to confirm the theories of particle physics.

A scientist pores over the photographic trails from the CERN bubble chamber in 1983.

Both the bubble and spark chambers worked on similar principles to the older cloud chamber. The detectors generally use liquid hydrogen, stored at its boiling point. although it is reported that Donald Glaser used beer in early versions of his bubble chamber. Just as the particles are entering the chamber, the pressure inside is lowered suddenly, causing the liquid to reach the very brink of boiling. High-energy particles leave a trail of tiny gas bubbles in this liquid, which are captured by cameras. The path that the particles take through a magnetic field applied to the liquid reveals their mass and charge. A spark chamber uses electrified gases flowing between metal plates to create sparks, which are captured in the same way. The biggest scalp of these detectors were the W and Z bosons, the force mediators of the weak interaction (the force that drives some radioactivity), which had been predicted in 1973 and observed as bubbles in 1983.

90 Ivy Mike: Another Big Bang

RESEARCH INTO THE BIG BANG THEORY AND HOW STARS BURN CALLED UPON NUCLEAR FUSION, a process by which small atoms merge together to make larger, heavier elements. In the 1950s, the power of fusion was demonstrated with H-bombs.

Where did all the elements come from? In the late 1940s, a theory of nucleosynthesis had been developed that said that all heavier elements were the product of nuclear fusion of hydrogen. Two hydrogen nuclei fused to form one of helium. In turn, helium could fuse into boron atoms, and so on. Generally, fusions produce unstable radioactive nuclei which break down into a stable form, representing one of the elements we find around us today. Fusion requires enormous forces, the kind you would find in the heart of a star, which is where most of the elements have been formed in the distant past. Fusion releases energy, too, and that is what makes stars shine. The heaviest elements—gold, uranium, and mercury, etc.—are born from fusion inside a supernovae, immense explosions that occur when giant stars die.

The Ivy Mike test detonation took place on Elugelab, a small island near Enewetak Atoll in the Pacific Ocean. The bomb was 450 times the power of the fission weapon dropped on Nagasaki in 1945. It completely destroyed Elugelab.

Edward Teller blows a bubble with colleague Hans Bethe. Teller likened the propagation of fusion reactions to bubbles spreading at the quantum level.

Hot war

In 1952, the US military revealed that they had harnessed the power of fusion when they detonated Ivy Mike, the first thermonuclear bomb. Also known as a hydrogen bomb, this weapon explodes due to fusion of tritium (a heavy isotope of hydrogen). The bomb had been designed by Edward Teller and Stanislaw Ulam. It used the heat of a fission bomb—like the ones used in World War II—to force the tritium to fuse, making an even bigger bang.

91 Masers, then Lasers

THE LASER IS AN ICONIC TECHNOLOGY, FOREVER LINKED WITH THE POPULAR IDEA OF SCIENTIFIC PROGRESS. However, we came close to this winning bit of optical equipment being known as a loser!

A laser is source of coherent, collimated light. Coherent means the light waves oscillate in time with each other. Collimated means the waves are parallel, rather than shining out in all directions. Laser light also contains a handful of wavelengths, perhaps just one, instead of a wide spectrum. All this means that laser light can be reflected, refracted, and focused very accurately without the beam diminishing.

The first laser was not a laser at all, but a maser: Microwave Amplification by Simulated Emission of Radiation. (Microwaves are high-frequency radio waves.) The principle is the same: A source of energy, generally light or electricity, is supplied to a "gain medium," making its electrons oscillate and emit a beam of radiation. The gain medium is often a crystal—sapphires are used to make lasers, for example. The maser was perfected in 1953, and by 1957 an optical version was developed. Strictly speaking, this development was best described as Light Oscillation by Stimulated Emission of Radiation, or *loser*. Thankfully that name never caught on.

Lasers are used in everything from removing unwanted hairs to reading DVDs and measuring the distance to the Moon—and of course in exciting light shows.

92 Neutrino Flavors

IN SOME RADIOACTIVE DECAY, PROTONS AND NEUTRONS CAN SWITCH FROM ONE TO ANOTHER, CHANGING THE ATOMIC NUMBER. When quantum physicists looked at this process they found that something was missing.

In alpha decay, the nucleus chucks out an alpha particle, made from two protons and two neutrons. The atomic number drops by two, and the mass number drops by four. Beta decay is more complicated. There are two forms. In the first, a neutron collapses into a proton. A neutron is ever so slightly heavier and is neutral. That extra mass and unneeded charge are released as an electron—the beta particle.

However, quantum theory states that as well as mass and charge, momentum and spin must be conserved, and that means there is a third particle in play. In the 1920s, this had been dubbed the neutron, but by 1932 that name had been given to the massive particle in the nucleus. The elusive particle is also neutral but has a tiny mass—too small to measure, even today—and so Enrico Fermi added an Italian twist, naming it the neutrino. Beta decay can also go the other way: A proton forms a neutron, this time releasing a positron and a neutrino. In fact, in the earlier example, the neutrino emitted is an antineutrino—a bit of antimatter popping out of the atom.

Electromagnetism has no effect on neutrinos, and gravity has an imperceptible impact, so these particles are hard to find. In 1956, the first antineutrino was detected indirectly by picking up its related particles. Beta decay produces electron "flavored" neutrinos. Muons are heavy versions of electrons, formed in particle collisions. In 1962, muon-flavored neutrinos were seen, and in 2000, tau-flavored neutrinos, associated with the short-lived tau particle, heavier still than the muon were spotted. In 2001, it was also discovered that neutrinos "oscillate"—change from one flavor to another.

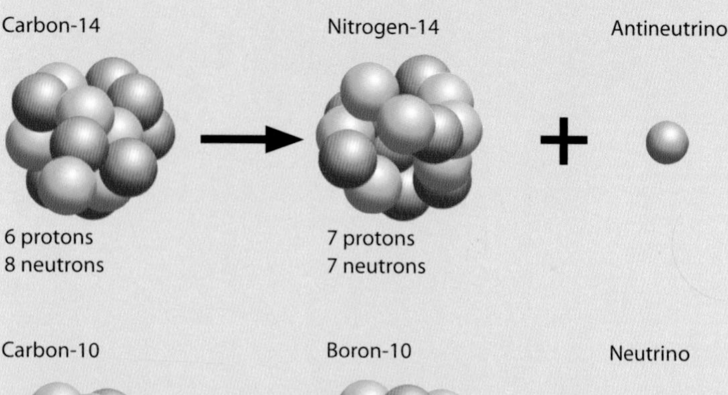
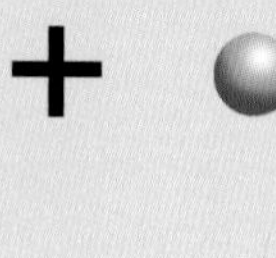
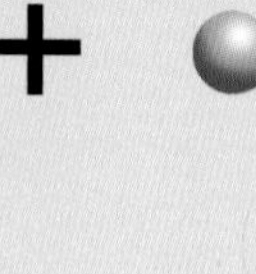

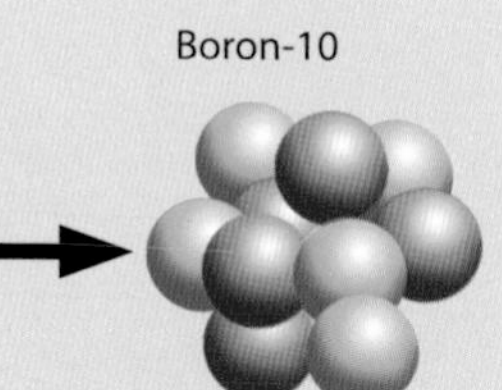
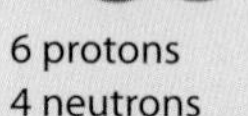

A diagram showing both kinds of beta decay. The neutrino is pushed out of the nucleus by the weak force, but after that it is only under the influence of gravity. Because neutrinos have an infinitessimal mass, gravity barely does anything to them.

Neutrino detectors are positioned deep underground to shield them from unwanted particle "pollution." This one is 2,000 m (6500 ft) underneath Canada. The sphere contains water, which gives out Cherenkov radiation when a neutrino hits. Huge numbers of neutrinos are streaming through Earth every second, but only a tiny percentage are picked up by even this giant detector.

93 Quarks: Strangeness and Charm

In 1964, particle physics went to the next level. It turned out that protons and neutrons were not elementary particles after all. They were actually made from a new gang of smaller entities.

Working independently of each other, two American physicists, Murray Gell-Mann and George Zweig, came to the conclusion that the massive particles, things like proton and neutrons, but also their smaller cousins, the mesons, were in fact made from still smaller particles. Gell-Mann named them *quarks*, from a word he'd read in a book by Irish writer James Joyce. Gell-Mann pronounces them *kworks*, but they are *kwarks* to most of us.

Chromodynamics is a theory describing the force between quarks inside protons or other large particles. It is an interaction between particles of varying "colors"—akin to charge but with six states rather than two. It all makes for a very vibrant picture.

Triples and doubles

Zweig and Gell-Mann proposed that neutrons and protons have three quarks. That made them the two most common members of a group of particles called the baryons. (Since that time a large number of other short-lived baryons have been seen fleetingly after high-energy collisions.) Mesons, such as kaons and pions, have two quarks. Together, baryons and mesons make up a larger group called the hadrons. It has been proposed that there may be supermassive hadrons containing four or five quarks, but so far these big guys have not shown up. It is not possible for one quark to exist alone.

Quark flavor

There are six types, or flavors, of quark—up, down, top, bottom, strange, and charm. The most stable and lightest are the up and down quarks; the others eventually decay into them. A proton has two ups and one down quark, while a neutron has one up and two downs. The exotic baryons contain other combinations and are very short lived. Some last just trillionths of a second. Each quark has an electrical charge, but this is either a third or two-thirds. Combinations of three quarks (making a hadron) always result in a whole number charge ranging from –1 to +2.

The mesons have one quark and one antiquark. They work in the mechanism that holds the nucleons—the protons and neutron particles in a nucleus—together. However, at the level of the quarks this force is mediated by yet more particles called the gluons. These are the bosons that are ultimately responsible for sticking matter together.

94 The Standard Model

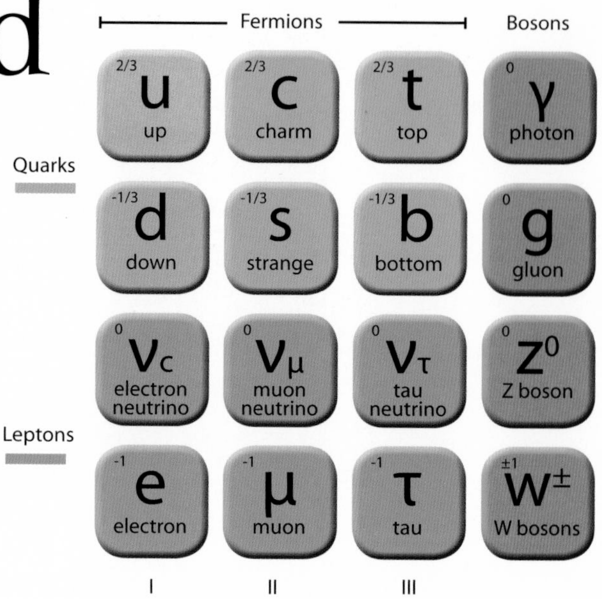

This chart shows the 17 confirmed particles in the Standard Model (remember there are two W bosons). The columns of fermions show the three "generations," with the most stable on the left becoming more short-lived to the right. And of course, each one of the particles above also has an antimatter version.

WAY BACK IN 1897, J.J. THOMSON'S DISCOVERY OF THE ELECTRON WAS THE FIRST STEP IN BUILDING THE STANDARD MODEL. This is a set of fundamental particles that took shape in the early 1970s with the aim of explaining every aspect of physics.

The Standard Model has a bold purpose and it has far to go in describing all forces and matter (dark matter, for example). However, it has been shown to be a very good way of understanding visible matter and three out of the four forces of nature.

There are 17 particles in the Standard Model today, probably 18 if we add the Higgs boson. Back in the early 1970s, there were lot of gaps, but the predictions held true, and over the years more and more complex detectors have picked off the missing particles.

Force and matter

There are four forces of nature. Gravity attracts masses over astronomical distances. It is the weakest force and so far has eluded being included in the Standard Model. Next comes electromagnetism, which creates attraction between opposite charges, and repulsion of like charges (or polarities when it comes to magnets). This force acts over

The ALICE detector at CERN, the leading European particle physics research center near Geneva, is built to watch large nuclei shatter as they are hit by a proton. The result is a quark–gluon plasma, a primeval soup of matter in its most basic form.

macroscale distances. The weak force acts only in spaces smaller than an atom. It is involved in pushing particles out of radioactive nuclei. Finally, the strong interaction keeps the atomic nucleus together and is the strongest, but shortest-acting force of all.

The Standard Model divides its particles into fermions, the particles that are acted upon by forces (and are subject to Pauli's exclusion principle), and bosons, the particles that carry the forces between the fermions. The fermions are further divided into six quarks (three with 1/3 charge and three with 2/3 charge) and six leptons (three neutrino flavors and their three associated charged particles, including the electron). The quarks are subject to the strong interaction, while the leptons are not.

Five bosons have been identified so far. The strong interaction is mediated by the gluon. Electromagnetism is carried by the photon. The weak force has three bosons. The W bosons (there are two, each with an opposite charge) act to push heavier leptons, like electrons, out of the nucleus in beta decay. The Z boson does it for the lighter neutrinos. (In alpha decay, it is the strong force and electromagnetism that does the pushing.) The missing feature is gravity, a force that acts on all the fermions without exception. It is postulated that the boson for gravity would be the graviton, but it is proving to be such a puny particle that no one can find it.

Investigating the Standard Model requires the largest and most expensive laboratories in the world, such as the Compact Muon Solenoid particle detector in the Large Hadron Collider, a huge particle accelerator in Switzerland.

95 String Theory

ALTHOUGH GREAT STRIDES HAD BEEN ACHIEVED IN THE FIRST HALF OF THE 20TH CENTURY, modern physics had become split into two reference frames: Quantum theory tackled the Universe at the atomic scale and below, while relativity dealt with everything else from specks of matter to entire galaxies. Could these two branches be entwined into a Theory of Everything?

The quick answer is not yet. Albert Einstein had spent the last 40 years of his career in search of a link between the quantum and the macroscale worlds. He failed to find one, and the search for a theory of everything continues to this day. The big schism between the two theories is gravity. The three other forces of nature could all be explained in terms of the Standard Model. However, gravity could only be truly understood on the astronomical scales of relativity.

In the late 1960s, string theory, actually more a family of theories, took a mathematical approach to the problem. The theory uses a number of techniques to try and tie up quantum waveforms with the space warps of relativity. It is inspired by topology, a field of bendy geometry, where angles and lengths are of no consequence—the important things are the connections.

Quantum mechanics represents subatomic particles as zero-dimensional points, but string theory treats them as one-dimensional lines, or strings. The particle's quantum property, such as spin, charge, or flavor, is represented by oscillations in the string. However, for that to work the string has to do more than wobble up and down and from side to side; it needs to oscillate in a number of other dimensions—the latest total is 11. These may be compact dimensions, only existing at the quantum level, or they may be universal spatial dimensions that our 3-D brains can barely conceive of. String theory is hard to test, although one of its predictions is superparticles, which connect all other particles into a symmetrical "everything." The world's most powerful particle physics labs now have superparticles in their sights. Will they find them?

String theory invites us to imagine that matter is comprised of a seething tangle of vibrating strings.

96 Hawking Radiation

BLACK HOLES ARE DIFFICULT TO SEE. THE FIRST EVIDENCE FOR THEM DID NOT SURFACE UNTIL 1971. Soon after, a young English physicist found he could use a black hole to study relativity and quantum physics at the same time.

Stephen Hawking's black hole radiation has never been observed but it predicts that black holes "evaporate" away.

As far back as the 18th century, the French polymath Pierre Simon Laplace proposed a *corps obscure*, a body with gravity so strong that its escape velocity was greater than light speed. In modern terms, that meant even light cannot get out of its gravity well. The result would be a literal black hole in space.

In 1915, Karl Schwarzschild calculated how small a mass had to be to have an escape velocity of the speed of light. This is its Schwarzschild radius. Earth's mass would have to fill a 9-mm (0.35-inch) dot to achieve such a gravitational pull. Actual black holes are even denser than this. They are formed when a giant star collapses into a speck. Such a massive body warps space-time in accordance with relativity, but being tiny it also falls under the purview of quantum theory.

You might think that anything that crosses the "event horizon" and falls into a black hole can never come out. However, in 1974 Stephen Hawking suggested otherwise. Quantum uncertainty tells us that within the shortest possible time period (10^{-43} seconds) virtual particles of matter and antimatter exist everywhere, ceaselessly forming and then annihilating each other. Opposing pairs of virtual particles on either side of an event horizon would be instantly separated when one is pulled into the black hole. The other particle that is released has become known as Hawking radiation. This shows that even black holes lose mass.

Black holes pull in all matter around them, sweeping space clean. As the material swirls toward the event horizon it heats up, offering evidence of the otherwise invisible body.

97 Spintronics

JUST WHEN YOU THOUGHT PHYSICS HAD GOT COMPLICATED ENOUGH, in 1988 a new branch opened up. It was a way of harnessing quantum effects and had a suitably modern name: Spintronics.

As its name suggests, spintronics is focused on controlling the spin of electrons for use in ever tinier electronics. Its foundations were laid with the discovery of giant magnetoresistance (GMR) in 1988. This mouthful is an effect seen in very thin (not giant at all) conductors. GMR has an impact when a conductor just a few nanometers thick is sandwiched between two other layers. In the top layer the electrons are free to spin in any direction. In the lower layer they cannot change their spin. The spinning electrons in the top layer interfere with those in the bottom layer, and that makes it impossible for an electric current (more electrons) to move through the middle layer. In other words there is "giant" resistance. A magnetic field will give order to the electron spins, and the resistance vanishes: GMR in action.

GMR is in action in the hard disk of a computer. The files are stored as a magnetic code on the disk. GMR is used inside the "head" that reads code as the disk spins. When the head passes magnetized regions, GMR will cause blips of current to pass through the head.

98 Dark Energy

FRITZ ZWICKY, THE DISCOVERER OF DARK MATTER, WAS ALSO A LEADING SUPERNOVA SLEUTH. In 1998, the largest survey of the superbright signatures of these exploding stars revealed another dark truth.

A shadow was cast over everything we knew about space and matter at the end of the 20th century. For 70 years, all the evidence suggested that the Universe had been expanding ever since the Big Bang, and the mutual attraction provided by gravity was slowly applying the brakes. The question remained whether there was enough matter—including the dark stuff—to bring the expansion to a halt. Or would the Universe get too enormous for gravity to prevent it from expanding forever.

Dark matter survey

The unknown factor was dark matter. Gravity is proportional to mass, so if astronomers could measure how fast gravity was slowing the expansion, it would give a clear idea of how much dark matter was out there. Astronomers began to survey type 1a supernovae. A supernova is a very large explosion that occurs in stars above a certain size—always much bigger than the mass of our own Sun. A 1a supernova reaches this mass and explodes in a very particular way. It forms in a binary system of a regular star and a white dwarf (the glowing core of a dead star). Material from the larger neighbor is pulled over to the white dwarf until its mass reaches the minimum supernova size—then boom! Type 1a stars are all the same mass and so have a standard brightness. That means they can be used to measure distances—the fainter ones are farther away than the brighter ones. The redshifts of their light also shows how fast the stars are moving away. The theory predicted that the more distant (and older) light would show a greater redshift than the nearer (and younger) sources. Older light would reveal an older expansion rate than younger sources. The differences could be used to calculate the effect of gravity on the expansion.

Then came the shock: The expansion of the Universe is not slowing; it is actually speeding up! The force of gravity is not pulling everything to a halt because there is another, wholly unknown force pushing matter apart. This force was named "dark energy," and physicists are still unsure what it is. Its effects only become manifest in the vast nothingness of space-time—and appear to get stronger as the Universe expands into ever more "nothingness." Eventually (in many billions of years), the dark energy will be so powerful that it will even pull atoms apart, creating an infinitely thin soup of subatomic particles.

Dark energy has changed our view of the Universe completely. It is estimated that dark energy comprises three-quarters of the Universe, while dark matter makes up most of the remaining quarter. Things like stars and atoms barely account for one percent of the whole.

99 The Hunt for the Higgs

Peter Higgs is the biggest name behind the new boson—currently named for him. However, five other 1960s' physicists have some claim on it. Perhaps one day its name will reflect this, but as is often the case, this name will probably stick.

UNTIL 2012, THE STANDARD MODEL LACKED SOMETHING RATHER CRUCIAL. It did not have a mechanism for giving matter its mass. The largest experiment ever conducted was underway beneath the fields of Switzerland to find it.

Mass is the property of some—but not all—subatomic particles. Specifically, it is the quarks that make up things like protons and neutrons, and the leptons, which include electrons and neutrinos. It is an easy leap to equate mass with weight, but what the property of mass really does, is it makes matter subject to the forces of nature. Weight is the pull of gravity, the weakest force of nature. Matter is also affected by the strong interaction, which keeps the atomic nucleus together; the weak force, which is involved in radioactive decay, and electromagnetism, the "opposites attract" force, which binds electrons to the atom, among other things.

All these forces are mediated by other particles called bosons, which move energy between masses. In 1964, Peter Higgs (and other researchers) suggested that a boson is also responsible for imparting mass to matter. A "Higgs field" created by these bosons became active in the very

The LHC is 27 kilometers (17 miles) long, cost about $6 billion, and took ten years to build. It fires two beams, less than 1-mm across, around the supercooled tube, before smashing them together inside the particle detector.

early Universe as pure energy was converted to the matter that now fills the Universe. Between 2008 and 2011, this idea was tested at the Large Hadron Collider (LHC) at CERN, an underground research center in Switzerland. The LHC, the most powerful particle accelerator ever built, smashes protons together at nearly the speed of light, to recreate the intense energy that existed just after the Big Bang. By 2012, CERN scientists had found evidence of a Higgs field forming during those collisions. The theory was correct, although the nature of the particle, the Higgs boson (or bosons), is still unclear, because it exists for less than a billion-billionths of a second.

100 Supersymmetry?

WHAT NEXT FOR PHYSICS? FOR A LONG TIME THE GREAT HOPE FOR A FUTURE BREAKTHROUGH HAS BEEN THE THEORY OF SUPERSYMMETRY. However, the latest research suggests that physicists may need to think again.

A symmetrical system remains the same when it is transformed in some way. To a physicist, symmetry is all about conserving quantum properties of a system. The theory of supersymmetry proposes that every boson is linked with a fermion, forming a pair of sparticles, or superpartners. (A fermion has a boson sparticle, while bosons have a fermion sparticle.) Superpartners share a set of properties (it is not yet clear what exactly). Whatever one particle does, its sparticle stays symmetrical. However, a fermion always has a half-spin to the boson's whole-number spin. One version of the theory suggests that sparticles are much heavier than their partners and could account for dark matter. Others suggest that the sparticles reside in "superspace" outside of the 3-D space we can perceive. The LHC is currently looking for sparticles, now that it has found the Higgs boson. Early results show no evidence of them. One commentator reported, "Supersymmetry may not be dead but these latest results have certainly put it into hospital." The search continues.

Engineers install some of the LHC's 450,000 optical fibers. Plans are afoot to upgrade the detector into the High Luminosity LHC, capable of bigger and brighter collisions by 2020.

101 Physics: **the basics**

SO WHAT DOES ALL THIS DISCOVERY ADD UP TO? Physics shines a light into every corner of the Universe—and even deals with phenomena that can never be observed. Here is a roundup of the basics for budding boffins.

What is energy?

We might as well start with a big question. Energy is what makes things happen in the Universe. Everything always has at least some energy—mass itself is a form of energy. It cannot be created or destroyed; the amount of energy now is the same as at

TYPES OF ENERGY

Acoustic
Carried by waves of pressure that travel through a medium. The ear detects acoustic energy as sound.

Radiant
Energy carried by light and waves of other forms of electromagnetic radiation, such as gamma rays and infrared.

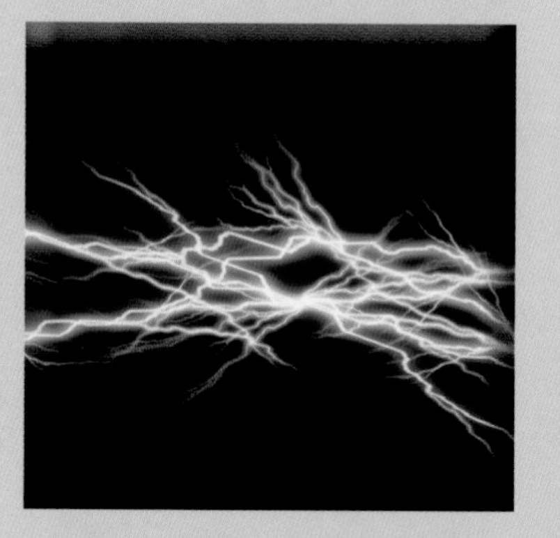

Electrical
The energy transferred as electrons or other charged particles form currents running through materials.

Thermal
The energy that moves atoms and molecules inside material. A high thermal energy makes the material hot.

the beginning of time. However, energy can be transferred between objects and can be converted into a number of different forms. This flow of energy is what makes the Universe dynamic, constantly changing, never still for long.

Mass and force

What is force?

A force is something that transfers energy from one object to another, and in the process alters it in some way, either changing its speed, changing its direction, or deforming its shape. There are four forces of nature: The strong force holds the atomic nucleus together; the weak force is involved in radioactivity; electromagnetism makes opposite charges attract each other, while like ones repel. Finally, gravity is a force of attraction that acts between all masses. The strong and weak forces only act inside the atom, and we all feel gravity as weight. When we push against another object, we are employing the electromagnetic force: The negative charge of the electrons that shroud the atoms of your hand are repelling the same charge around the atoms in the object. Instead of merging, the atoms force themselves apart.

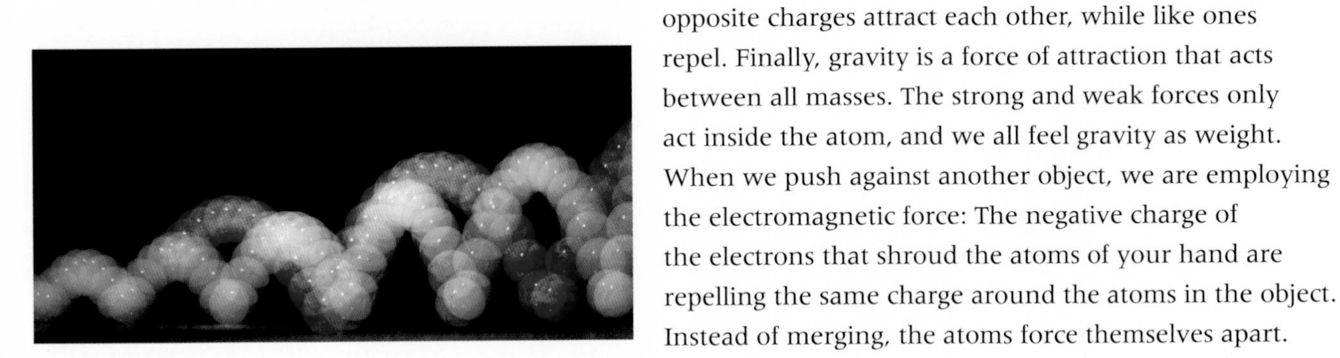

The bouncing balls follow these curving paths because a sideways force has set them in motion, but the downward force of gravity also keeps pulling them down.

What is mass?

All objects have mass, which is best defined as a resistance to forces. An object with a lot of mass requires a bigger force to move it than a smaller mass. This property of mass is called inertia, which ensures that a mass will maintain its motion (or lack of motion) until a force acts to change it. In physics the energy transferred when a force is applied to a mass is known as "work." Work is calculated as force multiplied by distance.

Kinetic
The energy of motion; a collision of objects will transfer kinetic energy between them.

Chemical
The energy given out or taken in during chemical reactions as chemical bonds are broken and new ones formed.

Nuclear
The energy stored at the heart of atoms in the nucleus. It is released in fission and fusion reactions and radioactive decay.

Potential
The water at the top of the falls has potential energy. This is transformed into motion as it falls to the bottom.

Motion

Turning a corner is a change in velocity, even if the speed remains constant. In the case of these motorbikes, a force is being applied by their engines which results in a sideways acceleration that changes the velocity to a new direction.

Speed and acceleration

Physics is very precise about the nature of motion. Speed is the distance an object travels every unit of time. Velocity measures the same thing as speed but it also takes into account the direction of the object. So two bodies traveling toward each other at the same speed have twice the velocity relative to each other.

An object will maintain a constant velocity until a force is applied to it. In the real world, the drag force of air or other friction will always bring an object to a stop, unless forces are continually applied in some way. However, in the vacuum of space, bodies

keep on moving, even without forces being applied. Applying a force will result in an acceleration, which is measured as the rate of change of velocity. A large force will produce a larger acceleration than a smaller one would.

Newton's cradle was designed to show conservation of momentum in action. As the red ball collides with the first silver one, it slows to a halt, transferring its momentum to that ball. That first ball does not move, but transfers its momentum through its neighbors to the ball at the end—which is the one free to move.

Momentum

An object in motion keeps on moving because of its momentum. Momentum is a measure of motion, calculated as mass times velocity. (The larger the mass and the faster it moves, the greater the momentum.) To stop a moving body, you must remove its momentum and that requires energy delivered by a force. Once the momentum has been reduced to zero, the body stops. However, momentum is not destroyed, merely transferred to the object (a hand, bat, or wall, for example) that delivered the force. The law of conservation of momentum is a primary piece

Peregrine falcons, the fastest animals on the planet, harness the pull of gravity as they swoop in for the kill.

of evidence that energy is not created or destroyed, just transferred between masses.

The pull of gravity

On or near to the surface of Earth, the force of gravity is a component in every motion. Gravity is a force that attracts masses together. It acts over huge distances between stars and even galaxies, but on Earth, our experience of it is dominated by the planet beneath our feet. A pull of gravity exists between every object and planet Earth. Obviously, the mass of the planet is incredibly large in comparison to even the biggest objects we can make, and so the force of gravity between it and us results in only an infinitesimal (so small as to be almost zero) motion of Earth. On the other hand, the force will result in a considerable change of the motion in the other body—a swooping bird, skydiver, or meteor—making it accelerate toward the ground.

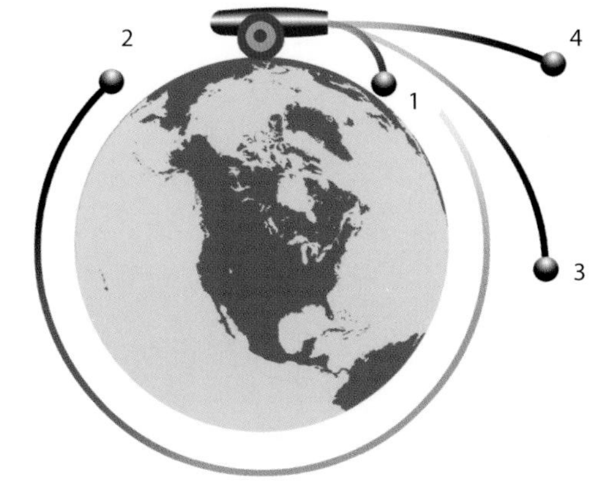

This diagram is borrowed from a similar one sketched in Isaac Newton's Principia. *It shows a hypothetical cannon that fires ball 1 below orbital velocity, so it follows a parabolic path to the ground. Ball 2 has an orbital velocity, ball 3 is at escape velocity and follows a large parabola away from Earth. Ball 4 is fired at above escape velocity so it follows a curve called a hyperbola up and away from the planet.*

Orbital forces

Gravity is also keeping the Moon in orbit around Earth—and the myriad artificial satellites that swarm around our world. How can a force that pulls things down result in a circular motion? The answer is that an orbiting object has been accelerated to a large horizontal speed relative to the surface of Earth. If the vertical component of its motion (the pull of gravity) outweighs this horizontal component, then the body will eventually plummet to the ground. If the horizontal component outweighs the pull of gravity, the body will fly away—or escape—from the planet. It has achieved "escape velocity." Orbital velocity is when the horizontal velocity and vertical velocity are in balance, and so the body just keeps on going around and around at the same altitude. It has achieved orbit.

Weight and mass measure up the same way on Earth. However, on the Moon, where the gravity is less, mass remains constant but weight is reduced.

Weight versus mass

Gravity is what creates weight, the force that pulls a body to the ground. While mass stays constant wherever you are, the weight depends on the pull of gravity. For example, the Moon has smaller mass than Earth and its pull of gravity is one-sixth as strong. A 60-kg, moon-walking astronaut has not lost any mass, but he or she will weigh only 10-kg on the Moon.

Waves

Waves are vibrations that transfer energy from one place to another. Some waves, like sound waves or the seismic waves of earthquakes, need a carrier medium, while others, like beams of light, do not. There are two main types of wave: Transverse waves (like light) oscillate up and down, while longitudinal waves (like sound) move as areas of compression and rarefaction (expansion).

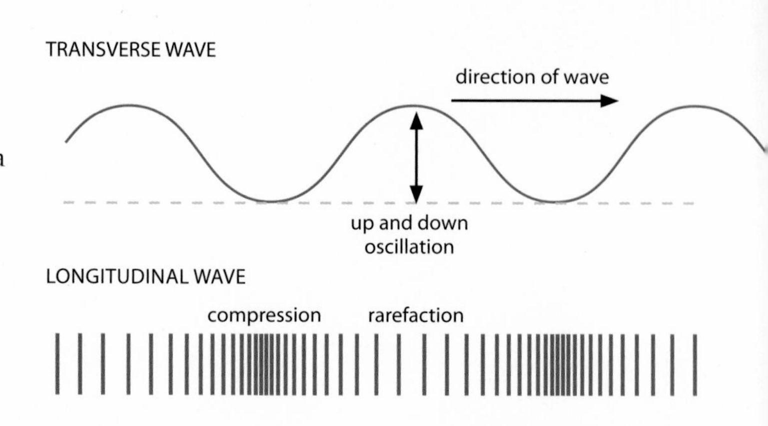

Measuring waves

Waves can be defined by wavelength, frequency, speed, and amplitude. The wavelength is the distance that the wave travels as it undergoes one oscillation, or single vibration. The frequency is the number of wavelengths completed over time. The speed is how quickly the wave can travel. When a wave has a fixed speed (as does light), its wavelength is inversely proportional to its frequency. Finally, amplitude is a measure of the size of the oscillation.

Below: The wavelength (λ) is constant wherever it is measured on the wave.

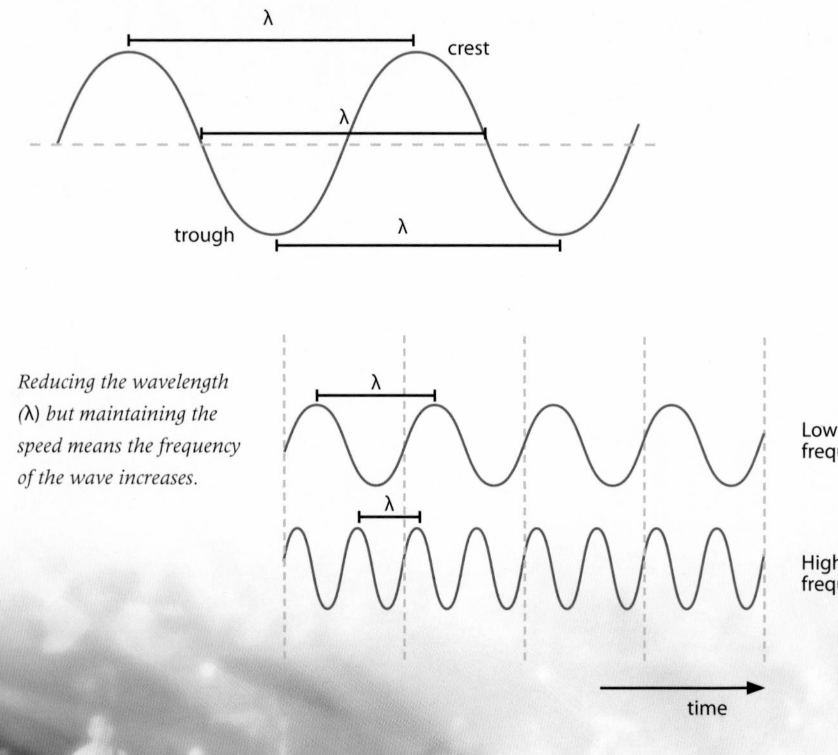

Reducing the wavelength (λ) but maintaining the speed means the frequency of the wave increases.

If all other aspects of a wave stay constant, the amplitude is a measure of the wave's power (energy per unit of time). A good example is that of sound: A loud sound wave has a high amplitude.

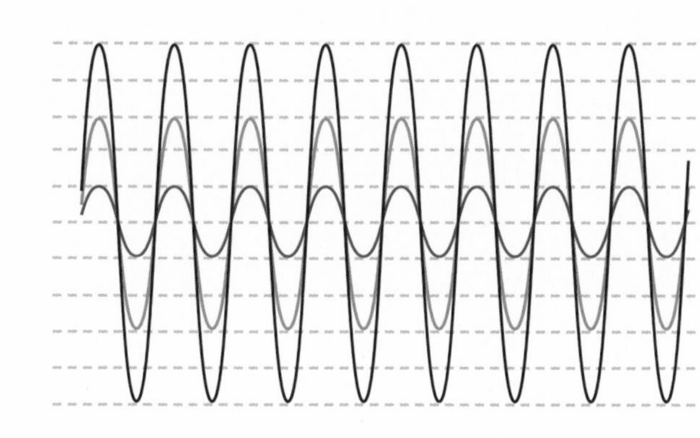

Much of our perception is based on waves. The sounds and sights of a concert reach the audience as a series of waves propagating from the stage.

Optics

Reflection and refraction

Optics is the study of how rays of light behave. The law of reflection states that the incident (arriving) angle of a ray is the same as the reflected angle. This ensures that light from a source is not jumbled, and means we can see reflections in smooth surfaces. When light refracts, rays are diverted by a specific angle. A lens can focus rays because light hits its curved surface at slightly varying angles. Therefore, their angles of refraction are also slightly different and the rays converge on one point.

The focal distance of a lens depends on its curvature. A higher curvature will bring the focal point closer to the lens.

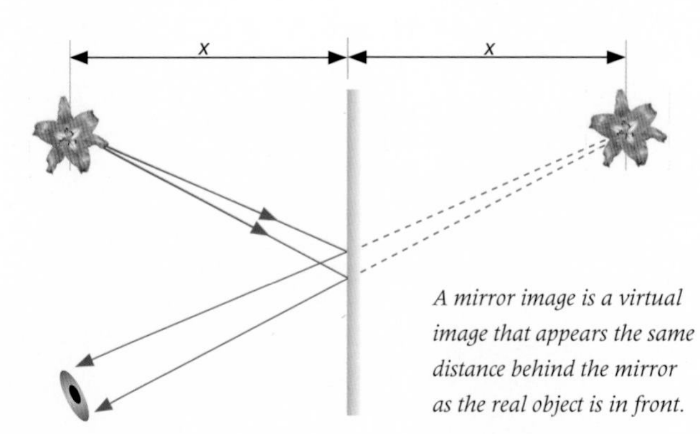

A mirror image is a virtual image that appears the same distance behind the mirror as the real object is in front.

Below: Reflection and refraction at work.

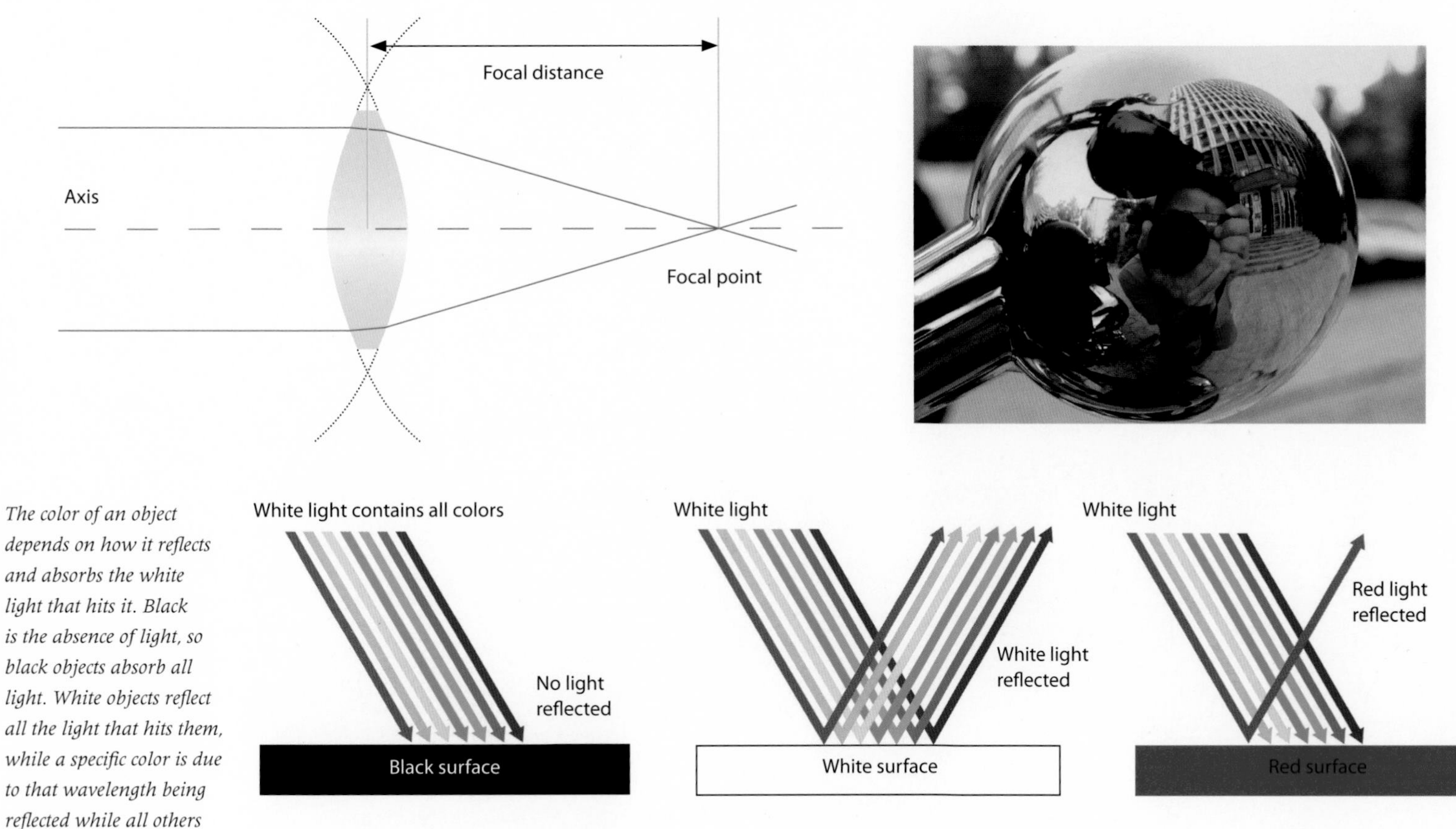

The color of an object depends on how it reflects and absorbs the white light that hits it. Black is the absence of light, so black objects absorb all light. White objects reflect all the light that hits them, while a specific color is due to that wavelength being reflected while all others are absorbed.

Interference

Like all waves, rays of light will interfere with each other. This is a result of waves having a different phase—their oscillations are out of time with other waves. When two waves are in phase (oscillating in time), constructive interference occurs, resulting in the waves merging into a single, more powerful wave. When the waves are out of phase, destructive interference results: The two waves cancel each other out.

Constructive interface

Crests and troughs merge

Destructive interface

Crests and troughs cancel

Electromagnetism

Since the 1820s, it has been known that electricity and magnetism are two aspects of one phenomenon called electromagnetism. It is based on the force that keeps negatively charged electrons bonded to the positively charged nucleus of an atom. However, this same force can be harnessed to produce electric currents, flows of electrons or other charged particles. Electric currents travel through materials called conductors, which contain electrons that are free to move. Materials that lack this property are called insulators. They do not allow electric currents through. An electric force field surrounds a conductor, making the current move. At the same time a magnetic field is formed, turning the conductor into a magnet as long as it is electrified.

Ohm's law

This diagram shows how Ohm's law can be used to relate current, voltage, resistance, and power, and how each one can be calculated by knowing some of the others.

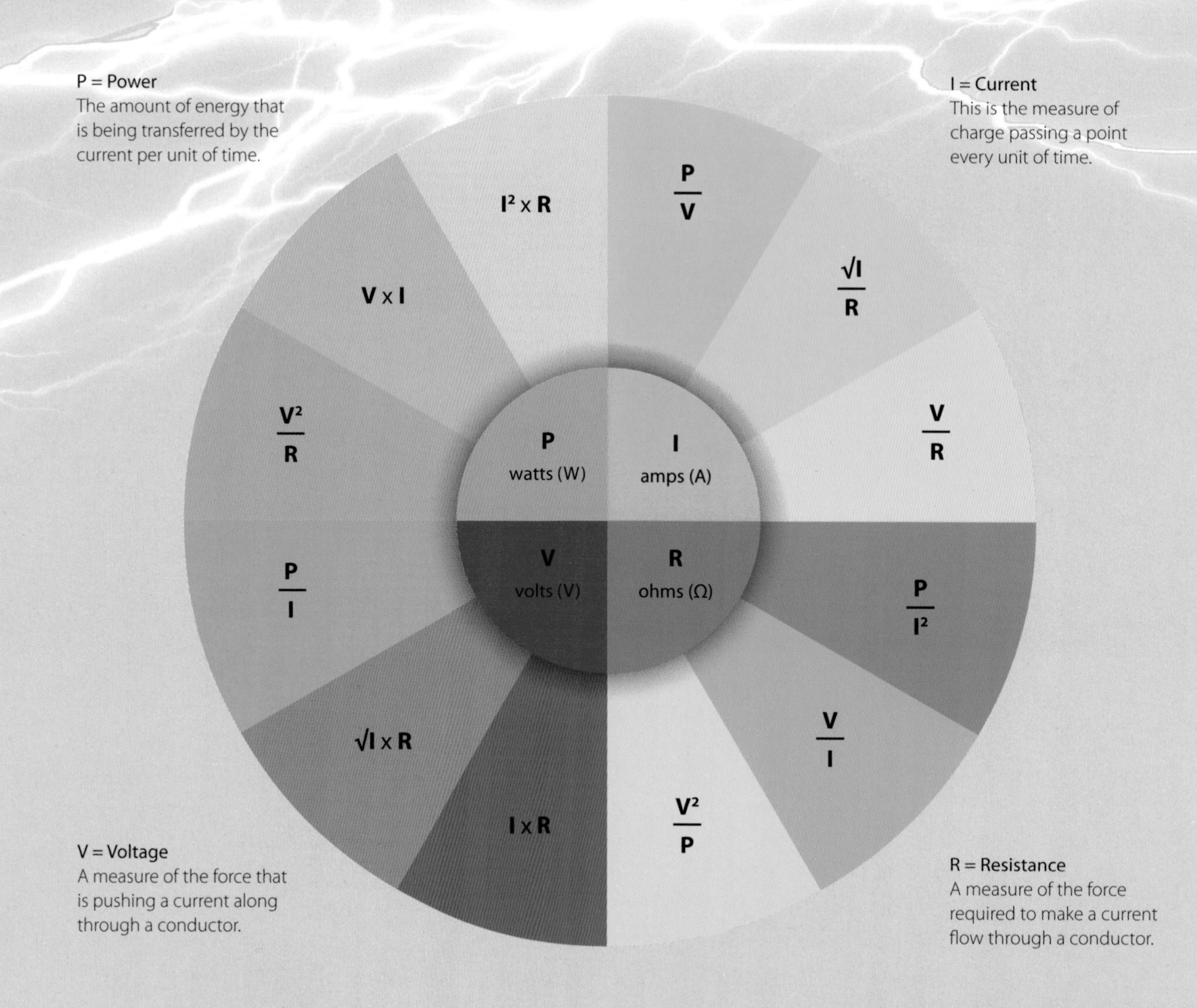

Magnetic field

A force field surrounds a magnet and can be imagined as lines of force connecting the north (N) and south (S) poles. Where the lines of force are most tightly packed, such as at the poles and near to the magnet, the force field is strongest. Farther away, as the lines of force thin out, the influence of the magnet diminishes.

Like electrical forces, magnetism obeys the "opposites attract" rule. The north pole of a magnet will attract the south pole of another, while like poles repel each other. This aspect is indicated with arrows on the lines of force. When arrows point the same way, the magnets attract, while opposite directions indicate repulsion.

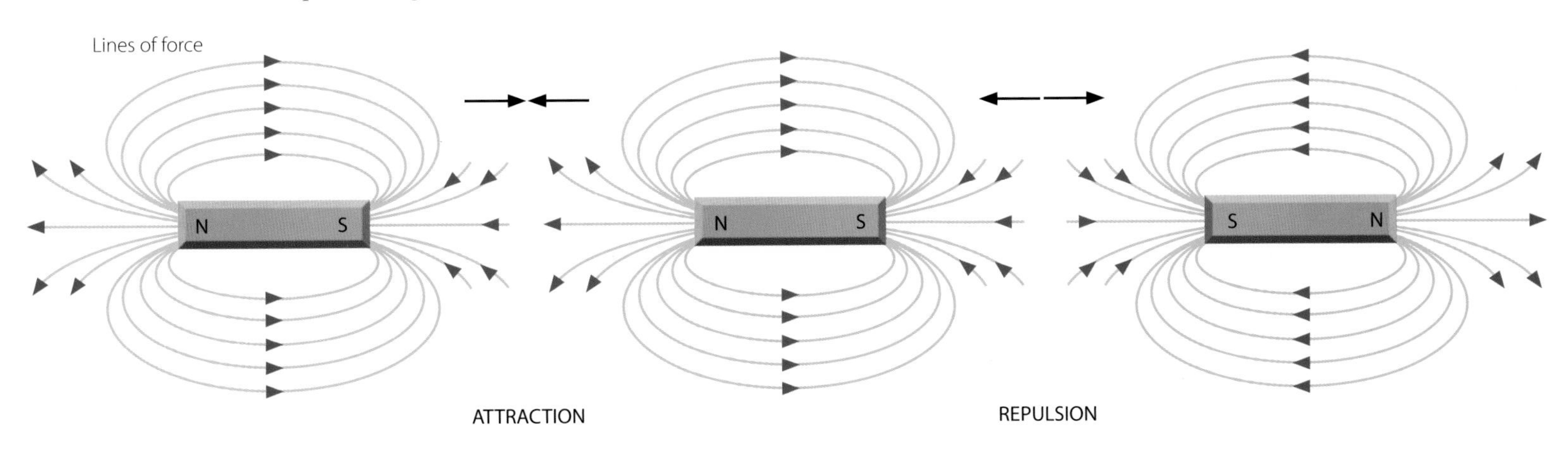

PHYSICS TABLE OF FORMULAS

Quantity	Description	Formula
Current	voltage / resistance	$I = \frac{V}{R}$
Voltage	current x resistance	$V = IR$
Resistance	voltage / current	$R = \frac{V}{I}$
Power	work / time	$P = \frac{W}{t}$
Time	distance / velocity	$t = \frac{d}{v}$
Distance	velocity x time	$d = vt$
Velocity	displacement / time	$v = \frac{s}{t}$
Acceleration	final velocity – initial velocity / time	$a = \frac{v_2 - v_1}{t}$

Quantity	Description	Formula
Force	mass x acceleration	$F = ma$
Momentum	mass x velocity	$p = mv$
Pressure	force / area	$P = \frac{F}{A}$
Density	mass / volume	$\rho = \frac{m}{v}$
Volume	mass / density	$V = \frac{m}{\rho}$
Mass	velocity x density	$m = V\rho$
Kinetic energy	1/2 mass x square of velocity	$E_k = \frac{1}{2}mv^2$
Weight	mass x acceleration due to gravity	$Wt = mg$
Work done	force x displacement in direction of force	$W = Fs$

IMPONDERABLES

THE PHYSICISTS OF THE LATE 19TH CENTURY BELIEVED THAT SCIENCE HAD MORE OR LESS EXPLAINED THE UNIVERSE. They were soon proven very wrong! Modern physicists are more humble as they tackle today's imponderables, thorny questions in need of answers.

Is human consciousness quantum physics?

Where doctors, biologists, and psychologists have failed, could physics provide the answer? The human brain has billions of cells which form trillions of connections. At the moment the human brain is not capable of understanding itself, but it has a few ideas. The human brain (and perhaps those of other big-headed creatures) has made a leap from a central processing unit for the body to a conscious mind full of abstract, imaginary thoughts. One suggestion is that this step up is a product of the wacky world of quantum mechanics. The basis for such an argument is that a brain can do what a computer cannot and can go beyond the limits of the axioms and assumptions that form the foundations of its activity. To do that it is suggested that quantum uncertainty is somehow harnessed to "compute" beyond the limits of the problem, program, or question being actioned. This is what gives rise to "original" thoughts. In itself this theory is little more than an original thought—we are no nearer to proving it. Perhaps we shall have to build a quantum computer and ask it to solve the problem for us.

How does gravity work on the quantum scale?

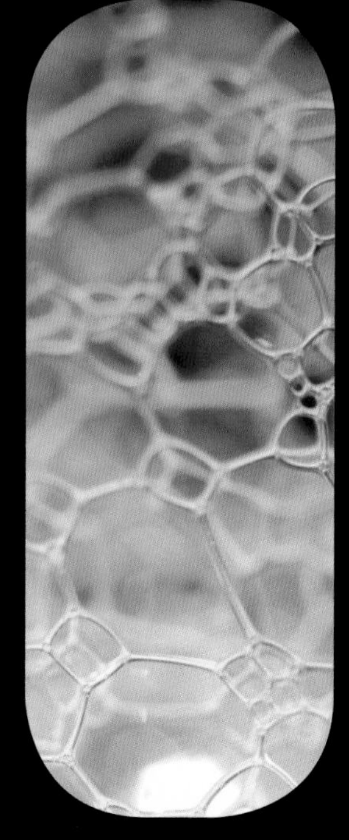

Look closely. The very nature of space and time might be a foam.

Gravity is all about space and time. Einstein's general theory of relativity shows how masses bend space and time, and gravity is the result of this warped geometry. This theory works great for the big stuff, like galaxies, stars, and apples. But how do we relate gravity to all the other forces of nature, which are explained at quantum level at the other end of the physics scale. Way down at this scale of subatomic particles, gravitational effects are too tiny to be observed over the push and pull of the other forces. Our only hope of a physics theory that explains "everything" is to explain gravity in quantum terms. One candidate to do just that is loop quantum gravity. This has the idea that space-time is atomic. In the same way that matter is formed from indivisible units—originally thought to be atoms, but now seen as quarks—loop theory says that space is made of "spin networks" measuring 1 planck length each. (Let's just say that is very small—in fact getting any smaller is impossible.) At the level of its spin networks, space-time is not warped by mass, and that allows gravity working on the large scale to be represented in the quantum world alongside the other forces of nature. Over time, the action of all the forces turns the spin network into churning "spin foam."

Can we shed light on dark flow?

Many observations have shown that a whole stream of galaxy clusters (many billions of stars) appears to be all heading in one direction, instead of spreading out in all directions. And they are traveling faster than they should. This phenomena was dubbed "dark flow" and it was suggested that the galaxies were being pulled toward an immense object beyond the edge of the visible Universe—perhaps even a whole other universe that was once connected to ours in the first flashes of the Big Bang. Putting it another way, space-time is on a tilt, and the contents of space are all sliding down it. A tilt of that magnitude could only be caused by a universe-sized mass. The most recent snapshot of the Universe made by the Planck space observatory seems to suggest that dark flow is the result of errors in earlier surveys, but the question is still open.

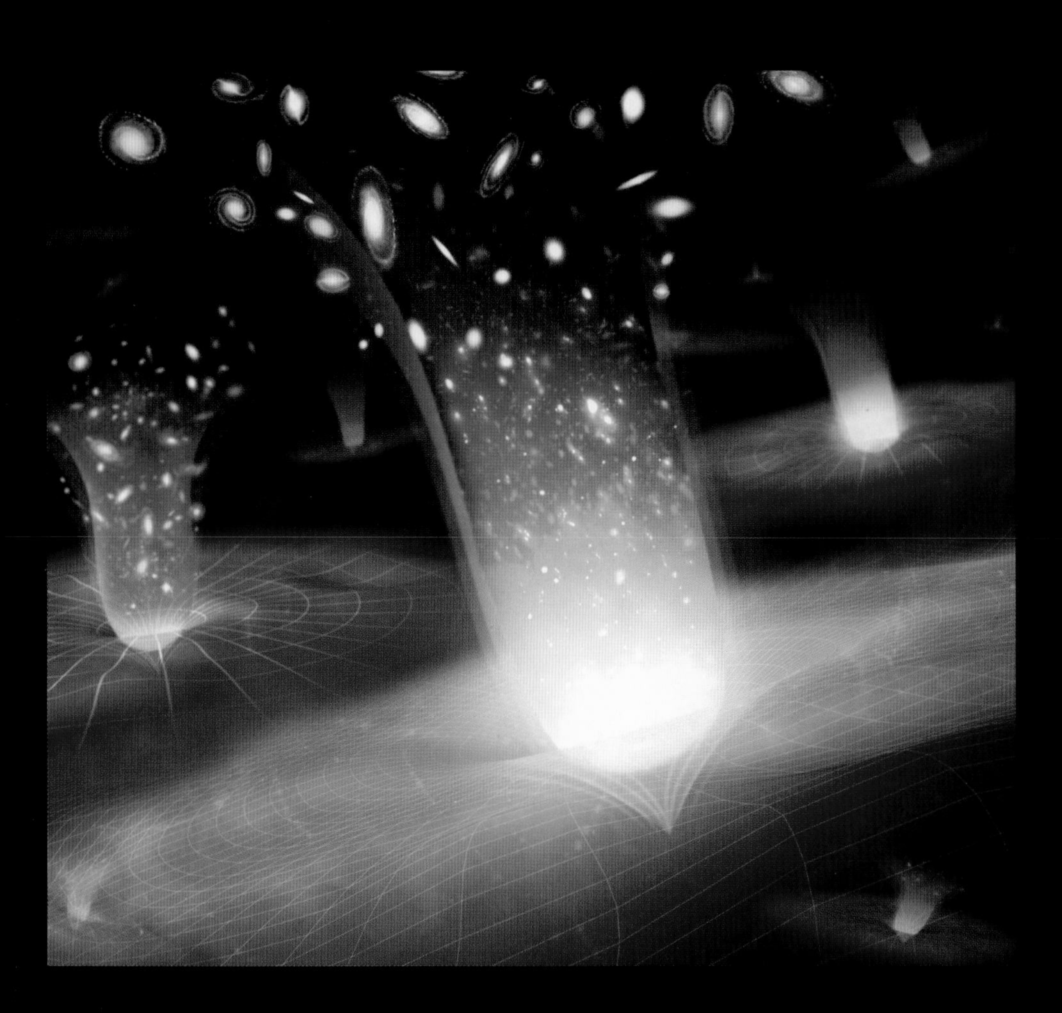

Is time always one way?

The passing of time is best described as an increase in entropy. Entropy is a measure of disorder. A system with high entropy is highly disordered, and the entropy of the Universe has always been going up—since the beginning of time. This is summed up in the second law of thermodynamics. Put another way it means that, seen as a whole, the matter in the Universe is forever getting colder and more dislocated and dispersed. In the big picture, the process never goes the other way, and is summed up as the "arrow of time." However, during radioactive decay, the particles involved appear to travel the other way in time—at least for a while. Is this all part of general thermodynamics, or is there a way to turn back time?

IMPONDERABLES

Is the dark energy an illusion?

Since 1998, there has been a new mystery in physics—dark energy, a mysterious force that appears to be accelerating the expansion of the Universe. However, is the ever-faster expansion of the Universe just an illusion? For the expansion to be speeding up, we are assuming that the Universe is a homogenous blob that expands in all directions equally. However, what if the Universe is a more complicated shape, where bulges and bubbles all expand at different rates. If our patch of space is in a particularly fast-moving region, it will appear that everywhere else is dashing away from us faster than it really is. Perhaps the true expansion rate of the Universe is slowing down, just as the Big Bang theory originally predicted. If so, that would do away with dark energy altogether, and once we correct the motion of the Universe, who knows where it will be heading?

The shape of the Universe is an open question. We could be getting a warped view of it from our galactic neighborhood.

Could a Universe exist without life?

The laws of the Universe are just right for life, for you and for me. If they were adjusted even slightly, stable stars, planets, and complex chemical-based life forms would become impossible. For example, the electromagnetic force is 10^{39} times stronger than gravity. Therefore, when the two forces (and others) work together to make stars ignite and emit heat and light, the results are stars that burn their fuels at a rate slow enough for stable planetary systems to form. Once all that cataclysmic bombardment and vulcanism is over, the rocky planets solidify and become part of a chilled-out solar system, where life is at least possible. (The latest estimates are that there are more planets than stars in the Universe!) If the proportions of the forces of nature were different, stars may never even form or would burn out too fast for the planets to mature—perhaps they wouldn't make enough metallic elements to form the heavy, rocky planets that life as we know it needs. So the next question is whether the Universe we observe is the way it is because it has to be like that for us to exist and to observe it? Put another way, can universes exist with laws of physics that preclude life? From the point of view of quantum physics, the answer is no. Without an observer such a Universe would remain a foam of unrealizable possibilities but contain nothing real at all. Then again ...

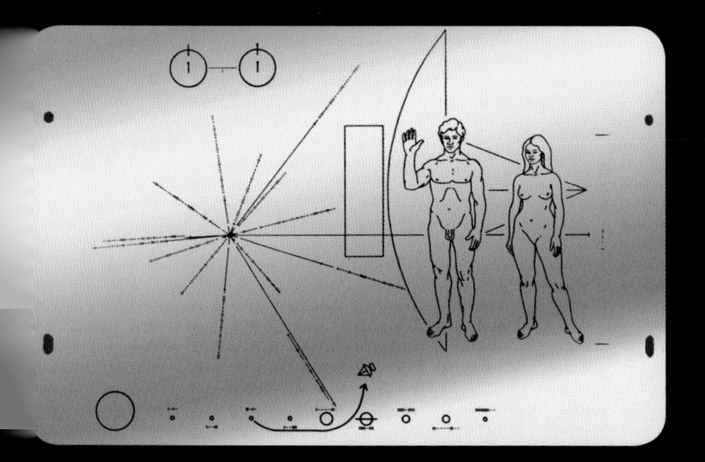

Is space filled with sterile neutrinos?

One of the general theories about dark matter, the missing material that makes up the majority of the Universe, is that it is comprised of WIMPS—or weakly interacting massive particles. One of the properties of mass is that it responds to force—you can push and pull it around. However, the theoretical WIMPS are unaffected by most of the forces of nature, meaning they could be all around us but they leave no trace of their presence. The prime candidate WIMP is the puny sounding sterile neutrino, a tiny particle that is only affected by gravity, the weakest of all forces of nature. The force of gravity is proportional to the mass of an object, and while the mass of neutrinos has yet to be measured they are estimated to be 100,000 times lighter than electrons. Therefore, even gravity barely does anything to sterile neutrinos. Could it be that the Universe is swamped with

The Great Physicists

THE GREATEST FIGURES OF PHYSICS ARE THOSE WHO HAVE been able to puzzle out the fundamentals of how the Universe works. Very often the search took place inside the minds of great theorists, like Einstein, Newton, and Feynman, while other breakthoughs were made with a more hands-on approach, such as Henry Cavendish weighing Earth from inside a shed, Ernest Rutherford using gold foil to reveal the atomic nucleus, and Enrico Fermi splitting an atom on a racketball court. Here, we take a look at the lives behind these events and more. Entries are arranged in chronological order, showing which great physicists were at work at the same time.

Thales of Miletus

Born	c.624 BCE
Birthplace	Miletus, western Turkey
Died	c.547–546 BCE
Importance	"Father of science"

There are no records of Thales's life. It is thought he was taught by an Egyptian priest and some reports suggest he had spent time in Athens and perhaps farther afield. The most concrete fact we have about his existence is that he is said to have been able to predict solar eclipses, and did so in a year that the Lydians were at war with the Medes (both neighboring kingdoms). We know that a final battle between these two states was fought at Halys in 585 BCE and ended when "day was turned to night" by an eclipse. This is assumed to be the one predicted by Thales and was viewed as an omen to stop fighting by commanders on both sides.

Democritus

Born	c.460 BCE
Birthplace	A Greek colony in modern-day Turkey
Died	c.370 BCE
Importance	Developed atomic theory

Democritus was born in a Greek colony in what is now western Turkey, and is reported to have traveled very widely. His work was inspired by lessons with Egyptian mathematicians, the magi of Persia, and the astronomers of Babylon. One story attests that Democritus blinded himself in order to increase the focus of his thoughts. He appears to have had a humorous outlook, often joking his way through critiques of other philosophies. Others were less fond of him. Plato led a campaign, mercifully a failed one, to have all of Democritus's work burned.

Aristotle

Born	384 BCE
Birthplace	Stagira, northern Greece
Died	322 BCE
Importance	Key figure in early Western science

Aristotle was born into the aristocracy of Macedonia, a northern region of Greece. He was the son of the king's doctor and like his wealthy peers, he finished his education in Athens, as a pupil of Plato. History reflects that Aristotle superseded his master, and become the most influential of all Greek philosophers. His work dealt with physics, astronomy, biology, and logic and became "received wisdom" from Europe to Asia. He got a lot wrong, but our modern understanding of physics began when people questioned the Aristotelian view.

Archimedes

Born	c.290–280 BCE
Birthplace	Syracuse, Sicily
Died	212–211 BCE
Importance	Discovered buoyancy principle

Aside from his work on buoyancy and mathematics, Archimedes is remembered for his inventions. The devices he designed for the defense of Syracuse from the Romans during the Punic Wars have passed into legend. His "claw" was said to shake ships while a "heat ray" set ships on fire before they could threaten the city. However, when the Romans eventually took Syracuse, Archimedes appeared unphased. So engrossed was he in his latest math puzzle, he ignored the orders of one belligerent legionary—who then hacked the great thinker to death.

Al-Biruni

Born	978
Birthplace	Khwarazm, Uzbekistan
Died	1048
Importance	Helps to unify field of mechanics

Many Islamic scholars built on the works of Classical Greece. Al-Biruni, who spoke seven languages and hailed from the eastern end of the Islamic world—he spent many years in what is now Afghanistan—also found inspiration in the science of India. His physics contributions were in mechanics and hydrodynamics, the motion of fluids. However, he is also remembered for calculating the radius (and circumference) of Earth by using a mountain peak (in what is now Pakistan) to form a huge right-angled triangle with the horizon and the center of Earth.

Al-Haytham

Born	c.965
Birthplace	Basra, Iraq
Died	c.1040
Importance	Founder of field of optics

In Europe of the Middle Ages, al-Haytham was known as simply as "The Physicist." He is perhaps the most prolific of the scientists from Islam's Golden Age. His hometown of Basra was a cultural hub in the 10th century, but al-Haytham finished his education at Baghdad's House of Wisdom, the top academic institution of the time. However, al-Haytham was not that wise. The story goes that his ill-fated move to Cairo was down to him boasting in Baghdad that he could control the Nile—a boast that landed him in deep water.

Averroes

Born	1126
Birthplace	Córdoba, Spain
Died	1198
Importance	Proposed an early conception of inertia

In the days of Averroes, Al Andalus, the caliphate in what is now Spain and Portugal (still reflected as Spain's southern region, Andalusia), was the most powerful place in the Islamic world. Averroes was born into one of its more powerful families, being the son of the chief judge. Like all self-respecting scholars, he did not just contribute to physics but is also remembered by medics and astronomers. His biggest impact was his philosophy. Averroism sought to merge religious truth with critical philosophy, a radical idea to this day.

Galileo

Born	February 15, 1564
Birthplace	Pisa, Italy
Died	January 8, 1642
Importance	Defined laws of fall and pendulums

So significant were the contributions of this scientist that he is known by his first name only. The son of a musician, Galileo Galilei chose a career in science, but was always on the lookout for a business opportunity—his family had frequent money troubles. The newly invented telescope was one such get-rich-quick scheme, earning him various pensions. However, the description of the Universe that he saw through his telescope put him in conflict with the Church, and to avoid jail and secure his income, Galileo was forced to recant.

Hooke, Robert

Born	July 18, 1635
Birthplace	Isle of Wight, England
Died	March 3, 1703
Importance	Hooke's law of elasticity

No true likeness of Robert Hooke survives, surprising for a figure who appears in so many different stories from 17th-century England. Hooke was a key figure in the foundation of the Royal Society of London in the 1660s along with the likes of Edmond Halley and Christopher Wren. That put him at the heart of many of the great leaps of the Scientific Revolution, including the investigation of gas by Robert Boyle, the discovery of the law of gravitation by Newton, and Huygen's use of oscillators to tell the time. Hooke was also one of the first scientists to turn a microscope on biological specimens. He reported seeing small enclosures within the tissue of plants, which he likened to the rooms of monks, naming them "cells."

Boyle, Robert

Born	January 25, 1627
Birthplace	Lismore, County Waterford, Ireland
Died	December 31, 1691
Importance	Discoverer of the first gas law

Robert Boyle's 1661 book, *The Sceptical Chymist*, was an early attempt to put the study of the elements on a scientific footing by questioning the mystical mumbo jumbo of alchemy. Aside from science, Boyle was very much a man of god, investing his fortune in the East India Company to spread Christianity (as well as provide a good return, no doubt). A bequest in his will was made to fund lectures on the latest religious thinking. Despite a few hiatuses, the Boyle Lectures have been held yearly ever since.

Newton, Isaac

Born	December 25, 1642
Birthplace	Woolsthorpe, Lincolnshire, England
Died	March 20, 1727
Importance	Formulated laws of gravitation and motion

Over and above his work on optics and motion, Newton's calculus made it possible to apply math to the ever-changing phenomena of nature. With a childhood marked by the loss of his father and rejection by his mother, Newton the man was secretive, selfish, and vindictive. Newton guarded discoveries so jealously that it was often decades before they were published. Much of the work on motion and math is reputed to have happened while he was at the family home in Lincolnshire, in retreat from the Black Death that was sweeping through the cities.

Franklin, Benjamin

Born	January 17, 1706
Birthplace	Boston, USA
Died	April 17, 1790
Importance	Devised concept of positive and negative charge

Before becoming one of the Founding Fathers of the United States and beginning a career as an ambassador and statesman, Benjamin Franklin had been an avid inventor and researcher. He is best remembered for work on electricity.

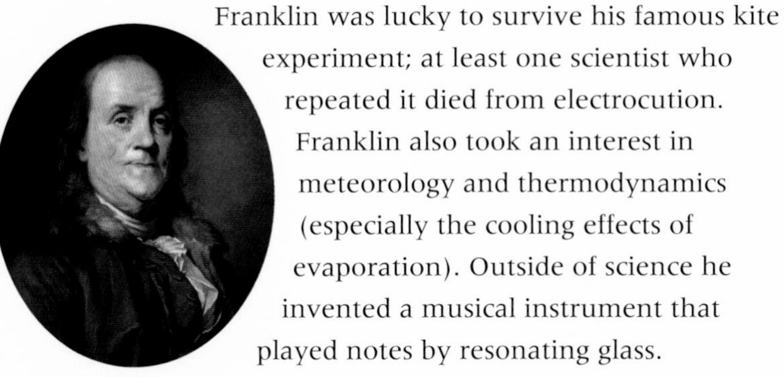

Franklin was lucky to survive his famous kite experiment; at least one scientist who repeated it died from electrocution. Franklin also took an interest in meteorology and thermodynamics (especially the cooling effects of evaporation). Outside of science he invented a musical instrument that played notes by resonating glass.

Cavendish, Henry

Born	October 10, 1731
Birthplace	Nice, France
Died	February 24, 1810
Importance	Measured gravitational constant

Henry Cavendish was born into scientific aristocracy. His father was a member of the Royal Society of London, soon to be joined by his son. A later cousin endowed Cambridge University with the Cavendish Laboratory, even today a leading research center. Cavendish was socially awkward. He worked in a laboratory built at the back of the family house, discovering hydrogen there in the 1760s. He lived a solitary life, leaving notes for staff. Nevertheless, he was a regular at Royal Society dinners, although he seldom spoke. As a result, many of his discoveries only came to light after he died.

Black, Joseph

Born	April 16, 1728
Birthplace	Bordeaux, France
Died	November 10, 1799
Importance	Discoverer of latent heat

Joseph Black opted for a career in medicine but his family were wine traders, and so as a child Black would have been exposed to the natural and artificial chemical processes employed by vintners. He maintained an interest in chemistry for his whole life, and this led to his discovery of "fixed air"—what we now call carbon dioxide—in the 1750s. This was an early step in the search for understanding the elements. Black was also a member of Scotland's literati of the day, meeting regularly with economist Adam Smith, philosopher David Hume, and engineer James Watt.

Volta, Alessandro

Born	February 18, 1745
Birthplace	Como, Italy
Died	March 5, 1827
Importance	Developed the electrical pile or battery

The first electrical device Volta worked on was the electrophorus. He did not invent this disk-shaped electrostatic generator but popularized it (even giving its name). His attention then turned to chemistry, which was a much more lively field than physics back in the late 18th century, but he began to combine the two. He formulated Volta's law of capacitance which states that charge and potential of a charged object are proportional. Napoleon Bonaparte had annexed Italy shortly before Volta developed his battery and got a personal demonstration. Volta became Count Volta soon after.

Dalton, John

Born	September 5 or 6, 1766
Birthplace	Eaglesfield, Cumberland, England
Died	July 27, 1844
Importance	Modern atomic theory

As a Quaker, John Dalton was barred from entering British universities, which did not approve of dissenting church sects. Dalton was mostly self taught and received informal schooling from John Gough, a natural philosopher from Manchester, England. (Gough also taught William Whewell, the man who later coined the word *science* itself.) Dalton lived frugally in Manchester even after election to the Royal Society of London. A unit of atomic mass is named dalton (Da) in his honor. One dalton is one-twelfth of the mass of an atom of carbon 12.

Henry, Joseph

Born	December 17, 1797
Birthplace	Albany, USA
Died	May 13, 1878
Importance	Codiscoverer of electromagnetic induction

History reflects that Joseph Henry discovered electromagnetic induction first, but the global reach of Faraday (not far behind) means it is largely seen as Faraday's discovery. Henry's work on applications of electromagnetism led to the development of the telegraph in the late 1830s, and he is the inventor of the electric doorbell, a simple electromechanical device that is found in homes across the globe. Henry also founded the National Institute for the Promotion of Science, which later became part of the Smithsonian Institution in 1846 and of which Henry served as the first secretary.

Faraday, Michael

Born	September 22, 1791
Birthplace	London, England
Died	August 25, 1867
Importance	Codiscoverer of electromagnetic induction

Born in poverty, Michael Faraday was apprenticed as a bookbinder. However, a visit to London's Royal Institution to hear the chemist Humphry Davy speak set his ambitions on a different track. His notes on the lecture so impressed Davy that the young Faraday was made his assistant. Faraday's own research achievements resulted in a conflict with his powerful mentor, which is thought to be one of the reasons that sent Faraday into a deep depression in middle age. By later life he was held in high esteem by the British people, but did little research.

Joule, James Prescott

Born	December 24, 1818
Birthplace	Salford, England
Died	October 11, 1889
Importance	Related heat to mechanical energy

James Joule was born next door to his father's brewery and he grew up to succeed him in the family business, working with heat and chemistry to produce a tasty product. Science was Joule's hobby—he had been fortunate to have been tutored by John Dalton in nearby Manchester. Hobby and work came together when Joule considered upgrading the brewery's steam engines with high-tech electric motors and he looked for a way of comparing the work done by both systems. Joule's grave is inscribed with 772.55, his measurement for the mechanical equivalent of heat.

Kelvin, Lord

Born	June 26, 1824
Birthplace	Belfast, Northern Ireland
Died	December 17, 1907
Importance	Calculated absolute zero temperature

Lord Kelvin started out as humble William Thomson. The son of a mathematics professor, Thomson was educated at a school within the University of Glasgow in Scotland, and it is little wonder he developed a wide set of interests in physics. As well as his work on thermodynamics, Thomson—ennobled to 1st Baron Kelvin in 1892—also worked on telegraphy and computing. He was involved with the first communications cable laid across the Atlantic and he invented an analog computer for predicting tides, which was so accurate that it was still in use in the 1970s.

Tesla, Nikola

Born	July 10, 1856
Birthplace	Smiljan, Croatia
Died	January 7, 1943
Importance	Developed electrical systems

Tesla is a hero in his homeland, or homelands. As an ethnic Serb the official Tesla museum and archive is in Belgrade. Being born in what is now Croatia, few Croat towns are without an *Ulica Nikole Tesle* (Tesla Street). However, Tesla was an Austrian citizen and spent most of his adult life in the United States. His father had wanted his son to be a priest, but as young Nikola lay ill from cholera, he promised to respect the boy's wishes to learn engineering if he recovered. Tesla moved to America in his 20s and worked for Thomas Edison and then his rival George Westinghouse, helping to develop electrical technologies.

Röntgen, Wilhelm

Born	March 27, 1845
Birthplace	Lennep, Prussia (now in Germany)
Died	February 10, 1923
Importance	Discovered X rays

We have this German physicist to thank for the fact that we are not sent for Röntgenograms when we visit the hospital. He preferred the simple term X ray, which is the one that stuck. Röntgen took out no patent on his imaging invention, making do with his academic's salary. When he was awarded the first-ever Nobel Prize for physics in 1900 he gave the prize pot away. He perhaps regretted this philanthropy when he became bankrupt in later life. Röntgen died of stomach cancer, most probably caused by a lifetime of exposure to dangerous radiation.

Thomson, J.J.

Born	December 18, 1856
Birthplace	Manchester, England
Died	August 30, 1940
Importance	Discovery of electrons

The founding father of particle physics, Joseph John Thomson revealed that atoms were not indivisible solids but constructed of yet smaller particles. An excellent school pupil, his parents planned for Thomson to be apprenticed as a steam-engine mechanic, but he was admitted to Trinity College, Cambridge, to read mathematics and physics at just 17 years old. He never really left, becoming the professor of physics there in 1884. Thomson's son George also won a Nobel prize for researching the wave-particle duality of electrons, the very particles discovered by his father.

Hertz, Heinrich

Born	February 22, 1857
Birthplace	Hamburg, Germany
Died	January 1, 1894
Importance	Discovers radio waves

Heinrich Hertz only lived to the age of 36—struck down by an autoimmune condition—but in that short life his legacy ensured that the unit of frequency, fundamental to all areas of science, in physics and beyond, was named for him. His development of the spark gap transmitter was used by Marconi and others to develop radio technology and led to television and today's wireless equipment. Hertz also discovered the photoelectric effect, which would later prove to be a springboard to understanding physics on the quantum level.

Curie, Marie and Pierre

Born	November 7, 1867 (Marie); May 15, 1859 (Pierre)
Birthplace (Marie)	Warsaw, Poland (then part of Russia)
Birthplace (Pierre)	Paris, France
Died	July 4, 1934 (Marie); April 19, 1906 (Pierre)
Importance	Pioneers in radioactivity

Marie Curie was born a Pole without a country. France offered an escape from oppression in her homeland, where it was even illegal to speak Polish. Marie moved to Paris and took two degrees at the Sorbonne, where she met Pierre. He had already made a discovery: Magnets lost their force above a critical temperature. Pierre was killed in a road accident at the height of his fame, and Marie filled his chair at the university, becoming France's first female professor of physics.

Planck, Max

Born	April 23, 1858
Birthplace	Kiel, Germany
Died	October 4, 1947
Importance	Originator of quantum physics

Max Planck, like Galileo before him, could have been a professional musician. He opted for a life of science, although he still entertained colleagues with his compositions. Planck's professor in Munich, Philipp von Jolly, questioned his career choices, saying that physics was dying: "Almost everything is already discovered, and all that remains is to fill a few holes." Planck proved him wrong although the work for which he is remembered took more than 20 years of hard labor. Planck's life was filled with tragedy. His first wife and two daughters all died young, while his son Erwin was executed for trying to overthrow Hitler.

Rutherford, Ernest

Born	August 30, 1871
Birthplace	Spring Grove, New Zealand
Died	October 19, 1937
Importance	Discovery of the atomic nucleus

Born on a humble farm on New Zealand's North Island, Rutherford began his academic career in Canada, but did his best work in Manchester and Cambridge, England. Later elevated to Baron Rutherford of Nelson, Ernest Rutherford's name appears throughout the early history of atomic physics. Chadwick, Geiger, Bohr, and Hahn all worked under him at some point, often directed by his theories toward their own personal discoveries. He was buried in London's Westminster Abbey near Isaac Newton. In 1997, element 104 was named rutherfordium (Rf) in his honor.

Meitner, Lise

Born	November 7, 1878
Birthplace	Vienna, Austria
Died	October 27, 1968
Importance	Codiscoverer of nuclear fission

Overlooked by the Nobel committee in favour of her German coworker Otto Hahn for the discovery of nuclear fission, Meitner nevertheless certainly left her mark on history in other ways. It was her work with Hahn that showed the tremendous destructive potential of a fission chain reaction and started the Manhattan Project to develop weapons that used fission. As a Jew, Meitner was forced to flee the Nazis, seeking refuge in Sweden. She spent the rest of her academic career there, before retiring to England. In 1997, element 109 was named meitnerium in her honor.

Bohr, Niels

Born	October 7, 1885
Birthplace	Copenhagen, Denmark
Died	November 18, 1962
Importance	Devised quantum atomic model

One of Niels Bohr's first positions was as goalkeeper in Akademisk Boldklub, a soccer team in Copenhagen. Bohr's academic career was just as vigorous. Bohr had reconfigured the model of the atom along quantum lines before his 28th birthday. By his 36th, he was director of his own institute of physics in Copenhagen. His actions in World War II precipitated a mass rescue of Danish Jews and after the war he was instrumental in the foundation of the International Atomic Energy Agency, set up to monitor nuclear technology.

Einstein, Albert

Born	March 14, 1879
Birthplace	Ulm, Württemberg, Germany
Died	April 18, 1955
Importance	Formulated theory of relativity

It is often said that Einstein was an average pupil (his handwriting was indeed terrible). However, from an early age he was already pursuing his own line of research. While still a teenager Albert was left to complete his studies in Munich while his parents sought work in Italy—he was not an attentive pupil. The poor academic record plagued his early career despite his talent. Einstein took a job as a patent clerk in Bern, Switzerland, in 1903. The untroubling job gave him time to work on the theories that would later propel him to the top of physics.

Lemaître, Georges

Born	July 17, 1894
Birthplace	Charleroi, Belgium
Died	June 20, 1966
Importance	Proposed an early version of the Big Bang theory

An unusual figure in the physics hall of fame, Georges Lemaître studied mathematics and physics while also preparing to take holy orders as a Catholic priest. This unusual career is all the more noteworthy because Lemaître was an early proponent of the expanding Universe and the Big Bang theory, which he described as "the primeval atom." His intuitive theory was ahead of its time, and it took a couple of decades for the evidence to catch up. In 2005, Lemaître was voted the 61st greatest Belgian of all time in a TV poll.

Fermi, Enrico

Born	September 29, 1901
Birthplace	Rome, Italy
Died	November 28, 1954
Importance	Set up first controlled fission chain reaction

Enrico Fermi's scientific prowess was born out of tragedy. His brother died young, and the teenager Enrico battled the grief through study. At the age of 24, Fermi became Italy's first professor of atomic physics. Within a decade he had opened the door to unlimited nuclear power. He went to Sweden to collect a Nobel prize in 1938, but did not go back to Rome. As a Jew in a Europe gripped by fascism, Fermi thought it better to continue his work into nuclear fission in the USA. Like many of his colleagues, Fermi died of cancer, unaware of the dangers of radioactivity.

Dirac, Paul

Born	August 8, 1902
Birthplace	Bristol, England
Died	October 20, 1984
Importance	Predicted existence of antimatter

Coming of age during post war austerity, the young Dirac lacked the funds to take a place at Cambridge. Instead, he had to make do with taking two degrees in engineering and mathematics in his home city. Eventually, he accrued enough scholarships to fund study away from home and within five years he had published the "Dirac equation," which opened up many new fields in quantum physics. Dirac's electron equation is regarded by some to be as significant a breakthrough as Einstein's theory of relativity.

Heisenberg, Werner

Born	December 5, 1901
Birthplace	Würzburg, Germany
Died	February 1, 1976
Importance	Proposed quantum uncertainty

Undoubtedly a key figure in physics, Werner Heisenberg's uncertainty principle is one of the first lessons in quantum mechanics. However, controversy surrounds this German scientist, who was a leading figure in the Uranium Club, the Nazi version of the Manhattan Project. Germany lacked the resources to develop atomic weapons in the time needed, but some suggest Heisenberg made deliberate errors to ensure slow progress. After the war he worked on peaceful nuclear technology and was the director of the Max Planck Institute for Physics.

Bethe, Hans

Born	July 2, 1906
Birthplace	Strasbourg, France (then Germany)
Died	March 6, 2005
Importance	Synthesis of elements in stars

One of many Europeans with Jewish heritage to move to the United States as fascism spread in the 1930s, Hans Bethe became the chief theoretician in the Manhattan Project. He also helped develop the hydrogen bomb, which harnessed nuclear fusion. His chief contribution among many was a theory of stellar nucleosynthesis, which showed how small atoms fused inside stars to make heavier elements. Ironically, the Alpher, Bethe, Gammov letter that set off the Big Bang theory was nothing to do with him. He was included because his name fitted.

Feynman, Richard

Born	May 11, 1918
Birthplace	New York City, USA
Died	February 15, 1988
Importance	Leading figure in quantum electrodynamics

With a twinkle in his eye and a knack for a good story, Richard Feynman was the most famous physicist of the second half of the 20th century. Not only did he lead many areas of particle physics, he was also an avid bongo player. While his first wife divorced him because all his mental calculus got in the way of their relationship, his third and final marriage was a better match. Richard and his wife Gweneth joked about visiting the Russian republic of Tuva, particularly because its capital, Kyzyl, had no vowels. However, Feynman died from cancer before he got there.

Gell-Mann, Murray

Born	September 15, 1929
Birthplace	New York City, USA
Died	–
Importance	Developed quark model

Murray Gell-Mann has worked at America's top schools: Yale, MIT, the Institute for Advanced Study, and Caltech. His contribution to physics was to impose some order on the seemingly chaotic hadrons—the "large" particles, such as protons and neutrons. Borrowing from Buddhism, his suggestion was the "eight-fold" path which led to the quark model (in collaboration with Kazuhiko Nishijima). Gell-Mann is also a founder of the Santa Fe Institute, which studies "complex adaptive systems," blending biology, economics, and linguistics.

Higgs, Peter

Born	May 29, 1929
Birthplace	Newcastle upon Tyne, England
Died	–
Importance	Proposes Higgs boson

Peter Higgs is now world famous for his link with the boson that gives mass to matter, which was discovered in 2012. Previously dubbed the God particle, Higgs's name is currently used for it, although there are moves to give it a name that reflects the contributions of others. Higgs is said to have been inspired to work in physics when he learned that Paul Dirac had attended the same school as him in Bristol. Most of Higgs's career has been spent at the University of Edinburgh, where he proposed his boson theory in 1964 —legend has it after a wet weekend hiking in the Scottish mountains.

Hawking, Stephen

Born	January 8, 1942
Birthplace	Oxford, England
Died	–
Importance	Discovered radiation from black holes

Trapped in a wheelchair due to a nerve disease that also robs him of natural speech, Stephen Hawking has become almost as much of a scientific icon as Albert Einstein, world famous as the brainiac who speaks through a computer. In 1979, he was made Lucasian Professor of Mathematics at Cambridge University, a post held by Newton and Dirac before him. Hawking's 1988 book, *A Brief History of Time*, became one of the best-selling popular science books in history.

BIBLIOGRAPHY AND OTHER RESOURCES

Books

Ananthaswamy, Anil. *The Edge of Physics*. Boston: Houghton Mifflin Harcourt, 2010.

Atkins, P.W. *Galileo's Finger: The Ten Great Ideas of Science*. Oxford: Oxford University Press, 2004.

Duhem, Pierre. *The History of Physics before Einstein*. New York: Encyclopedia Press, 1913.

Feynman, Richard. *Feynman Lectures on Physics*, Vols 1–3. Boston: Addison–Wesley, 1964.

Gribbin, John (ed.). *A Brief History of Science*. Lewes: Ivy Press Ltd, 1998.

Hawking, Stephen. *A Brief History of Time*. London: Bantam, 1988.

Lindley, David. *Uncertainty: Einstein, Heisenberg, Bohr, and the Struggle for the Soul of Science*. New York: Doubleday, 2007.

MacArdle, Meredith (ed.). *Scientists: Extraordinary People who Changed the World*. London: Basement Press, 2008.

Milhorn, H. Thomas. *The History of Physics*. College Station, Texas: VBW Publishing, 2010.

Parsons, Paul (ed.). *30-Second Theories*. Lewes: Fall River Press, 2009.

Simonyi, Károly. *A Cultural History of Physics*. Boca Raton: CRC Press, 2012 (Hungarian edition 1978).

Suplee, Curt. *Milestones of Science*. Washington, D.C.: National Geographic Society, 2000.

Museums

Canada Science and Technology Museum, Ottawa, Canada. www.sciencetech.technomuses.ca

China Science and Technology Museum, Beijing, China. www.cstm.org.cn

Cité des Sciences et de l'Industrie, Paris, France. www.cite-sciences.fr

Computer History Museum, Mountain View, California, USA. www.computerhistory.org

Copernicus Science Centre, Warsaw, Poland. www.kopernik.org.pl/en/

Deutsches Museum, Munich, Germany. www.deutsches-museum.de

Deutsches Technikmuseum/German Museum of Technology, Berlin, Germany. www.sdtb.de

Exploratorium, San Francisco, USA. www.exploratorium.edu

Franklin Institute Science Museum, Philadelphia, USA. www2.fi.edu

Huygensmuseum Hofwijck, Netherlands. www.hofwijck.nl

Kepler Museum, Prague, Czech Republic. www.keplerovomuzeum.cz

Massachusetts Institute of Technology Museum, Cambridge, Massachusetts, USA. www.web.mit.edu/museum

Museo Galileo, Institute and Museum of the History of Science, Florence, Italy. www.museogalileo.it

Museum of the History of Science, Oxford, UK. www.mhs.ox.ac.uk

Museum of Science and Industry, Chicago, USA. www.msichicago.org

Museum of Science and Industry, Manchester, UK. www.mosi.org.uk

MuseumsQuartier, Vienna, Austria. www.mqw.at

National Museum of Nature and Science, Tokyo, Japan. www.kahaku.go.jp/english

Norwegian Museum of Science and Technology, Oslo, Norway. www.tekniskmuseum.no

Observatory Museum, Stockholm, Sweden. www.observatoriet.kva.se/engelska

Pavilion of Knowledge, Lisbon, Portugal. www.pavconhecimento.pt/home

Powerhouse Museum, Sydney, Australia. www.powerhousemuseum.com

Science Museum, London, UK. www.sciencemuseum.org.uk

Shanghai Science and Technology Museum, Shanghai, China. www.sstm.org.cn

Smithsonian Institution, Washington DC, USA. www.si.edu

Archives, Individual Exhibits, and Preserved Equipment

Hans Bethe Papers, Cornell University Library, New York, USA. www.rmc.library.cornell.edu

Joseph Black Papers, Edinburgh University Library, Edinburgh, Scotland, UK. www.lib.ed.ac.uk

Niels Bohr Library and Archives, American Center for Physics, Washington DC, USA. www.acp.org

Maria Skłodowska-Curie Museum, Warsaw, Poland. www.muzeum-msc.pl

Paul Dirac Collection, Florida State University, Tallahassee, USA. www.fsu.edu

Albert Einstein Papers, Hebrew University of Jerusalem, Israel. www.huji.ac.il

Faraday Museum/Notebooks, Royal Institution of Great Britain, London, UK. www.rigb.org

Enrico Fermi Collection, University of Chicago, Chicago, USA. www.uchicago.edu

Richard Feynman Archives, Caltech, Pasadena, California, USA. www.caltech.edu

Murray Gell-Mann Papers, Caltech, Pasadena, California, USA. www.caltech.edu

Werner Heisenberg Papers, Max Planck Institute, Munich, Germany. www.mpp.mpg.de/english/index.html

Joseph Henry Papers, Smithsonian Institution, Washington DC, USA. www.siarchives.si.edu/history/exhibits/henry/henry-papers-database

Robert Hooke Folio, Royal Society, London, UK. www.royalsociety.org

James Prescott Joule Papers, University of Manchester Library, Manchester, UK. www.library.manchester.ac.uk

William Thomson, Lord Kelvin Collection, University of Glasgow Library, Glasgow, UK. www.special.lib.gla.ac.uk

Georges Lemaître Archive, Université catholique de Louvain, Louvain-la-Neuve, Belgium. www.uclouvain.be

Isaac Newton Papers, Cambridge University Library, Cambridge, UK. www. cudl.lib.cam.ac.uk

Rutherford Museum, McGill University, Montreal, Canada. www.physics.mcgill.ca/museum/rutherford_museum.htm

J.J. Thomson Papers, Trinity College Library, Cambridge University, Cambridge, UK. www.trin.cam.ac.uk

Websites

Nobel Foundation: www.nobelprize.org

Apps

Launchball, Science Museum for iPhone

LHSee for Android

Wolfram|Alpha for Android, iPad, iPhone

INDEX

Cataloging-in-Publication Data has been applied for and may be obtained from the Library of Congress.

ISBN 978-0-9853230-6-6

Series Concept and Direction: Jeanette Limondjian
Design: Bradbury and Williams
Editor: Meredith MacArdle
Proofreader: Marion Dent
Picture Research: Louise Thomas, www.cashou.com
Consultant: Dr. Mike Goldsmith
Cover Design: Jokooldesign

SHELTER HARBOR PRESS
603 West 115th Street Suite 163
New York, New York 10025

For sales in the UK and Europe, please contact info@worthpress.co.uk

Printed and bound in China by Imago.

10 9 8 7 6 5 4 3 2 1

PICTURE CREDITS

BOOK

Alamy/19th era 2 2-3 background; The Bridgeman Art Library Ltd. 134 bottom right; Mary Evans Picture Library 136 top left; Craig Hiller 81 top left; imagebroker 85 bottom right; INTERFOTO 4 center left, 26 right, 50 bottom left, 66; ITAR-TASS Photo Agency 139 top right; Gordon Langsbury 121 top left; MLaoPeople 120 center left; North Wind Picture Archives 68 bottom left, 70 bottom left; PHOTOTAKE Inc. 7 bottom, 82; Pictorial Press Ltd. 73 bottom right; The Print Collector 73 top left; Randsc 108 bottom right; RGB Ventures LLC dba SuperStock 117; sciencephotos 46 bottom, 81 bottom right; World History Archive 5 left, 16 top right, 29 bottom left, 35, 75 bottom right. **American Physical Society** 105 top left. **© CERN**/99 bottom, 110; Maximilien Brice 116 bottom; Claudia Marcelloni 111, 116 top right, 139 bottom left. **Corbis**/100 bottom left, 106 bottom left, 137 top left, 138 bottom right; Roger Antrobus 131 bottom right; Bettmann 5 center right, 20 bottom left, 33, 40, 42 bottom, 78, 84, 90, 92, 96 bottom right, 97, 99 top right, 137 bottom right, Endpapers; Stefano Bianchetti 74 bottom left; DK Limited 4 bottom left, 8 bottom left, 45 bottom right, 50 top right, 52 bottom right, 91 bottom right, 94 bottom right; Kevin Fleming 102 bl, 105 bottom right; Shelley Gazin 139 top left; Hulton-Deutsch Collection 72 bottom, 93; David Lees 26 bottom left; Ocean 63; Science Photo Library/Mehau Kulyk 129 top right; Sygma/Bureau L.A. Collection 43 top right; Tarker 49 top right; Underwood & Underwood 136 top right; Michael S. Yamashita 113 top left, 139 bottom right. **Dreamstime.com**/Empire331 48; Nickolayv 15 top center; Leon Rafael 24; Nico Smit 21; R.N. Whalley 22 bottom; Wollertz 131 bottom left. **Edgar Fahs Smith Image Collection, Rare Book and Manuscript Library, University of Pennsylvania** 44 bottom left, 45 top right, 52 top left, 58 center right, 60 top left, 87 bottom right, 133 bottom left, 134 top left, 134 bottom left, 135 top left, 135 bottom right, 136 bottom left, 137 top right, 138 top left. **Mary Evans Picture Library**/11 top right, 12 left, 12 bottom right, 30 top right, 49 bottom left, 69, 88 center right; Alinari Archives/Bruni Archive 101 bottom; Iberfoto/Fonollosa 130 bottom left; Imagno 138 bottom left; INTERFOTO/Sammlung Rauch 68 top right; SZ Photo 80, 138 top right. **from Harper's New Monthly Magazine, No. 231, August 1869**/ 55. **Ivan Kuzmich Federov, *Portrait of Mikhail Lomonosov*, 1844** 39 top right. **Image courtesy of the History of Science Collections, University of Oklahoma Libraries** 8 top, 19 top right, 23 top right, 27 bottom left, 30 bottom left, 30 bottom center, 32 top right, 32 bottom left, 34 center left, 39 bottom left, 54 bottom left, 59, 130 top right, 132 top, 132 bottom left, 132 bottom right. **iStockphoto.com**/John Butterfield 28 bottom left; Elemental Imaging 58 left; Hulton Archive 17 bottom left; Denis Kozlenko 41 bottom left, 133 top right. **Alfred Leitner** 100 top right. **Library of Congress, Washington, D.C.**/6 bottom left, 20 top right, 42 center left, 45 bottom center, 56 top right, 57, 67 center left, 83; Brady-Handy Photograph Collection 134 top right; George Grantham Bain Collection 62 top left, 71, 135 top right, 135 bottom left, 136 bottom right; Joseph-Siffrèse Duplessis 133 top left; Henry S. Sadd 37 bottom left; Jack Orren Turner 137 bottom right; Doris Ullmann 86 top right. **Albert Abraham Michelson** 65 center left. **NASA**/ AMES Research Center 129 bottom left; ESA and the Hubble Heritage Team (STScI/AURA) 86 bottom, 96 center left; ESA and Felix Mirabel (French Atomic Energy Commission and Institute for Astronomy and Space Physics/Conicet of Argentina) 113 bottom right; Johnson Space Center 75 center left; Ernst Mach 64 bottom left. **Sutton Nicholls, *Monument to the Great Fire of London*, c.1753** 29 bottom right. **Science & Society Picture Library**/Science Museum 41 bottom right, 53, 94 top right, 95, 103 bottom right. **Science Photo Library**/Mark Garlick 127 bottom left; Ted Kinsman 119 center left; Mehau Kulyk 2-3 center; New York Public Library 25, 62 bottom; David Parker 85 center right; Detlev van Ravenswaay 104; Volker Steger 65 bottom; Sheila Terry 61. **Shutterstock.com**/121 bottom left; argus 118 bottom center right, 124 background; Esteban de Armas 119 bottom center right; Binkski 8-9 center; Catmando 115; Andrea Danti 101 top right; Goran Djukanovic 122; Igor Golovnlov 131 top right; Alex Kalmbach 118 bottom left; koya979 87 top right; Peter G. Pereira 7 center right, 109; maisicon 112; Morphart Creation 17 center right, 18 top right, 28 top right, 58 bottom center; Maks Narodenko 32 bottom right; Werner Stoffberg 119 bottom left; Andrey Sukhachev 118 bottom right; weagle95 5 top right, 107; Rob Wilson 126 top right. **Thinkstock.com**/10 bottom left, 11 left, 14 top left, 23 bottom left; Dorling Kindersley RF 9 right, 72 top left, 85 top left, 131 top left; iStockphoto 36, 47 top left, 54 center, 77, 89, 114, 120 bottom right, 123, 126 bottom left; 128; Ryan McVay 60 center right; Photodisc 22 top right; Photos.com 4tr, 18 bottom left, 27 top right, 34 bottom, 37 top right, 38 top right, 43 top left, 56 bottom left, 67 bottom right, 74 top right, 130 bottom right, 133 bottom right. **U.S. Department of Energy** 9 top left, 106 top right. **U.S. Nuclear Regulatory Commission** 98. **U.S. Navy** Photographer's Mate 3rd Class, Jonathan Chandler 64 top. **Bartolomeu Velho, *Cosmographia*, 1568** 14 bottom right. **wallz.eu** 103 top left. **Roy Williams** 91 top left, 118 bottom center left, 119 bottom center left, bottom right, 127 top right.

TIMELINES

Alamy/INTERFOTO; The Print Collector. **© CERN**/Maximillien Brice. **Corbis**/Bettmann; Stefano Bianchetti; DK Limited; Rune Hellestad; Reuters/Sean Adair; Reuters/Dan Lampariello; Francis G. Mayer; Science Photo Library/Mehau Kulyk. **Dreamstime.com**/Jaroslav Bartos; Luis Manuel Tapia Bolivar. **Edgar Fahs Smith Image Collection, Rare Book and Manuscript Library, University of Pennsylvania**. **Mary Evans Picture Library**/AISA Media; Everett Collection; Glasshouse Images; Grosvenor Prints; Iberfoto/BeBa; INTERFOTO; INTERFOTO AGENTUR; INTERFOTO/Toni Schneiders; Pump Park Photography; Rue des Archives; SZ Photo. **Image courtesy of the History of Science Collections, University of Oklahoma Libraries**. **iStockphoto.com**/Karl-Friedrich Hohl. **Library of Congress, Washington, D.C.**/Brady-Handy Photograph Collection; Warren K. Leffler; Matson Photograph Collection. **Harriet Moore**. **NASA**/JHU/APL; MSFC; M. Weiss (Chandra X-Ray Center). **Science & Society Picture Library**/ Science Museum, London. **Shutterstock.com**/DVARG; Lludmila Gridna; Ale Justas; Atliketta Sangasaeng. **Thinkstock.com**/Dorling Kindersley RF; Hemera; iStockphoto; Photodisc; Photos.com. **U.S. Navy**/Ensign John Gay. **Roy Williams**.

Publisher's note: Every effort has been made to trace copyright holders and seek permission to use illustrative material. The publishers wish to apologize for any inadvertent errors or omissions and would be glad to rectify these in future editions.

Probl I

Investiganda est curva Linea ADB
qua grave a dato quovis puncto A
datum quodvis punctum B vi grav
suae citissime descendet

Solutio.

A dato puncto A ducatur recta in
et super eadem recta describatur t
(ductae et si opus est productae) occurr
ADC cujus basis et altitudo sit ad
spective ut AB ad AQ. Et haec
punctum B et erit Curva illa
ad punctum B vi gravitatis suae.